THE THREE QUESTIONS

Compelling Arguments for the Existence of God
and His Communication with Us

———————

Ing. Orlando Hernández

Always be prepared to offer an explanation to anyone who asks you to justify the hope that is in you. However, do so with gentleness and respect.

1 Peter 3:15-16

To all those who, through their participation in my conferences, have shown me that rational contributions are becoming increasingly important as pillars supporting the foundation of our faith.

And

To my wife Pilar who, with her presence, support, and help, accompanied me throughout this journey.

And

To my daughter Cata, so that she and those of her generation can recognize the truth and be amazed by it.

ARCHDIOCESE OF MIAMI
Office of the Archbishop

Decreto

THOMAS G. WENSKI

por la gracia de Dios y el favor de la Sede Apostólica
Arzobispo de Miami

El texto del libro *"Las Tres Preguntas"* ha sido revisado cuidadosamente y está exento de todo aquello contrario a la fe o la moral tal como lo enseña la Iglesia Católica Romana.

Por lo tanto, de acuerdo con el canon 824 del *Código de Derecho Canónico*, otorgo el *approbatio* necesario para la publicación del libro *"Las Tres Preguntas."*

Este *imprimatur* es una declaración oficial de que el texto está exento de error doctrinal o moral, y puede ser publicado. Esto no implica que la concesión de este sello signifique estar de acuerdo con el contenido, opinión o declaración expresadas por el autor del texto.

Dado en Miami, Florida, el diecinueve de agosto en el año de Nuestro Señor dos mil veintiuno.

Arzobispo de Miami

Attestatio et Nihil Obstat

Cancellarius

9401 Biscayne Boulevard, Miami Shores, Florida 33138
Telephone: 305-762-1233 Facsimile: 305-757-3947

Table of Contents

INTRODUCTION

I am neither a deacon, a theologian, nor a member of the Catholic Church's hierarchy. I am a dedicated layperson with a background in systems engineering and an MBA. This presentation introduces my first book, *What You Wanted to Know About the Catholic Church but Were Afraid to Ask*. I should clarify that I am neither an astronomer nor a biologist, as I have relied heavily on these fields—fascinating to me since childhood—to address the first question of the book: *Does God exist?*

I have always been fascinated by understanding mechanisms and how various systems operate, from simple devices like mousetraps to intricate workings of computers. Two fields in particular have captured my curiosity: astronomy and genetics. Throughout my life, numerous questions have arisen, seeking answers about the universe and life itself. How did everything come into existence? What is the origin of the matter that constitutes the universe? Why have we discovered many planets like Mars, Venus, or Jupiter beyond our solar system, but not one like Earth? Why do celestial bodies move in predictable patterns, enabling us to forecast the sky's appearance thousands of years into the future? How do stars generate their light? In biology, how does a fertilized egg develop into a human being? How does a single cell differentiate into various specialized cells such as heart, lung, brain, and bone cells? How can a tiny cell process complex information? What mechanisms allow our bodies to regenerate skin only where it is damaged, without affecting healthy tissue? Why does new skin form unique fingerprints on fingertips but not on palms? And how do cells know how to repair damaged tissues efficiently?

I had not expected that my interest in astronomy and genetics—initially motivated purely by intellectual curiosity—would evolve into the focus of my writing. Furthermore, I did not foresee that the sciences would become powerful allies in the quest for rational arguments supporting the existence of God. Louis Pasteur, the founder of microbiology, once remarked, "A little science distances us from God, but more science draws us nearer to Him." I can attest to the truth of this statement: God guided me to science, and in turn, science led me to God.

I must admit that when I first purchased my Bible (The Popular Bible, published by Herder), my purpose was not to explore spiritual teachings but to understand the biblical account of creation. My initial motivation was purely intellectual. I also found it fascinating how closely the biblical creation story aligns with the Big Bang theory[1]. How could Moses, the author of Genesis, have known that the universe had a beginning? How could he have described a light source on the first day[2] when luminous stars were only created on the fourth day? How could he have known that, according to the law of conservation of energy[3], all organic matter in living organisms originates from the soil? How could he have understood that life began in the oceans rather than on land, contrary to common observation? And how could he have known that, based on the law of biogenesis, only life can produce life?

In my first book, I explored thirty-three questions that I believe every Catholic should be able to answer confidently, without relying on typical responses like "This is an act of faith" or "This is what the Church teaches." Over time, I realized that three fundamental questions needed

---

[1]According to this theory, matter originated as an infinitely small point of extremely high density which, at a specific moment, exploded and began expanding in all directions. This event gave rise to what we now know as our universe, including both space and time.

[2]In 1978, the Nobel Prize in Physics was awarded to Penzias and Wilson for their discovery of cosmic microwave background radiation, which is attributed to the explosion of that original source of energy.

[3]It was discovered in the mid-19th century through the work of Mayer, Joule, Helmholtz, and others.

to be addressed first, as they form the basis of understanding and engaging with faith. These questions are: Does God exist? If so, does He communicate with us? And if He does communicate, can we trust that communication? These core questions echo the approach used by Saint Paul in his discussions with the Greeks, serving as the foundation for meaningful dialogue about faith and belief.

> Then Paul stood before them in the Areopagus and said: «Men of Athens, I have seen how religious you are. For as I walked around, looking carefully at your shrines, I noticed among them an altar with the inscription, 'To an Unknown God.' What, therefore, you worship as unknown, I now proclaim to you. The God who made the world and everything in it, the Lord of heaven and earth, does not dwell in shrines made by human hands. Nor is He served by human hands as though He were in need of anything. Rather, it is He who gives to everyone life and breath and all other things. From one ancestor, He created all peoples to occupy the entire earth, and He decreed their appointed times and the boundaries of where they would live. He did all this so that people might seek God in the hope that by groping for him they might find him, even though indeed He is not far from any one of us [...] God has overlooked the times of human ignorance, but now He commands people everywhere to repent, because He has fixed a day on which He will judge the world with justice by a man whom He has appointed. He has given public confirmation of this to all by raising him from the dead.». (Acts 17:22-31)

Saint Paul affirms, "The God who made the world and everything in it." This statement confirms that God, the Creator, exists and provides guidance on how we should live: "...and He decreed their appointed times and the boundaries of where they would live...". This indicates that God communicates with us, and the resurrection of Jesus demonstrates that he is the Messiah: "...He has given public confirmation of this to all by raising him from the dead." Therefore, we can trust in this communication. However, many Catholics do not possess sufficient rational arguments to confidently answer these fundamental questions, often assuming that everyone would naturally respond affirmatively.

Some time ago, the British Humanist Association[4] launched an advertising campaign across the UK, which later expanded throughout Europe[5]. The campaign's slogan was: "There's probably no God. Now stop worrying and enjoy your life." The word "probably" was included to protect the organization from potential legal actions by religious groups. They also published what they call the "bible for atheists," authored by their former president, British philosopher A.C. Grayling. This publication, titled *The Good Book: A Humanist Bible*[6], is structured as a collection of books like the Bible, including sections such as Genesis, Wisdom, Parables, Lamentations, Songs, Acts, Epistles, and Proverbs. A youth version, *The Young Atheist's Handbook*, was written by science teacher Alom Shaha. It aims to guide teenagers on how to live without relying on religion and is distributed free of charge in secondary schools across England and Wales by the association. Are Catholics prepared to address the presence of organizations like the British Humanist Association?

Currently, many children and society are misled by a simplistic and misleading form of atheism, which is often based on falsehoods and half-truths, primarily using Darwin's theory of evolution as the main argument. It is increasingly important to stay informed about recent scientific discoveries that challenge the foundations of naturalistic theories regarding the origins of the universe and life, thereby supporting the idea of a Creator as the source of all existence. Over fifty years ago, thanks to these discoveries, numerous scientists—including biologists, chemists, physicists, mathematicians, and paleontologists—express serious doubts about Darwin's original proposals Darwin's original proposals (the list of signatories to the well-known *Scientific Dissent from Darwinism*[7] exceeds thousands). These scientists reject the

---

[4]Founded in 1896, the Association promotes humanism and currently has over seven hundred thousand active members in nearly seventy cities across the UK. Its current president is comedian Shappi Khorsandi.

[5]In Spain, the campaign was funded by the Atheists of Catalonia and the Union of Atheists and Freethinkers associations.

[6]It was published in March 2011 by the publishing house Walker & Company.

[7]The scientists who signed the dissent statement come from around the world and hold degrees from prestigious universities such as Yale, Princeton, and Stanford. Each signatory was required to possess a Ph.D. in a scientific discipline or an M.D. degree

notion that chance alone drives the process of life's development and emphasize the importance of intelligence as the underlying cause of creation.

Does a 'ball' of energy mark the origin of the universe? Yes, but not in the way naturalists describe their complex theories. They believe that nothing caused the initial energy that led to everything. For us, God, as the creator and designer, brought it into existence; only He can create something from nothing. This 'ball' exploded, giving rise to time and space, but not by His own desire, as naturalists suggest, but as part of God's divine plan. The emergence of single-celled life on Earth was not a mere coincidence caused by an electrical discharge in a chemical-rich sea, but the result of divine intervention, infusing matter with the necessary information to form a cell capable of evolving into all known life forms. Regarding species evolution, it is not solely as naturalists claim—where one species mutates into another over time—but as the fossil record shows: species appeared suddenly and fully formed, with complex systems such as bones, nerves, and circulatory structures. After their appearance, they adapted to their environment through small modifications, known as microevolutions. We believe in theistic evolution. Most importantly, we are created with a purpose; nothing is created without a reason.

When discussing issues covered in this book with atheists, you may encounter individuals who search for impossible explanations, like the expression from Cervantes' *Don Quixote* about looking for three legs in a cat. Some people challenge logic and common sense. For example, imagine Charlie's mother comes home in the afternoon and finds some missing cookies. She notices crumbs on the floor and footprints of shoes leading to the cookie jar and then away from it, providing clear evidence of what happened.

---

while serving as a professor of medicine. The statement reads: "We are skeptical of claims about the ability of random mutation and natural selection to explain the complexity of life."

The full declaration, along with a detailed list of signatories, is available at: https://www.discovery.org/m/2019/02/A-Scientific-Dissent-from-Darwinism-List-020419.pdf.

For more information, you may also visit: https://dissentfromdarwin.org/

What evidence do we have?

1. Missing cookies.
2. Crumbs on the floor.
3. Footprints leading toward the cookie storage and away from it.

How can we explain the mysterious disappearance of the cookies?

1. Charlie ate them.
2. The older sister ate them.
3. Mom ate them.
4. Dad ate them.

If we consider additional evidence, such as the fact that the older sister left for a trip in the morning and does not like cookies, it becomes reasonable to dismiss possibility two. Furthermore, since the shoe footprint matches a small size rather than a large one typical of parents, possibilities three and four can also be ruled out. Based on this logical analysis, it is most likely that the only explanation for the missing cookies is possibility one. Do you agree?

Some individuals may challenge this conclusion, aligning with the type of characters Cervantes described in his writings. They might suggest alternative explanations for the disappearance of the cookies. Possibilities include the sister lying about her trip, waiting for the house to be empty, wearing her brother's shoes, and collecting the cookies, leaving crumbs behind. Others might argue that the father orchestrated the scene to make it seem as if Charlie was responsible or even propose that an alien was involved. These perspectives highlight the variety of interpretations that can arise in analyzing such situations.

These new theories proposed by some may indeed offer explanations for the missing cookies. But are they logical? Do they truly make sense? Are these alternatives genuinely worth considering? To those who put them forward, they might seem convincing. However, to the average person—who often relies on simpler reasoning—they may not be as persuasive. Ockham's Razor[8] is a methodological and philosophical principle that advocates simplicity when choosing between competing theories. It suggests that if two theories yield the same results, the simpler one is more likely to be correct. You may not have encountered this principle before, but it strikes you as reasonable and intuitive— because it aligns with what sound reasoning naturally dictates.

This book presents a series of logical and well-reasoned theses that explain events in a way that is both comprehensible and acceptable to reason. The explanations are grounded entirely in available evidence, deliberately avoiding absurdities that hinder the formation of clear and meaningful ideas.

Drawing on scientific discoveries in astronomy, physics, and biology, I address the profound question: "Does God exist?" Physics has progressively unveiled the intricacies of matter and energy, revealing a staggering number of "coincidences" that had to align with extraordinary precision for a life-supporting universe to emerge. These alignments go far beyond what could be attributed to random chance.

For much of modern science, the existence of the atom was taken as a foundational assumption. Today, however, we understand far more about the complex conditions necessary for the formation of even a single stable hydrogen atom. And once an atom comes into existence, progressing to more complex structures—such as stars—requires far more than mere chance. It demands coordination, planning, and flawless execution.

In short, it suggests the presence of design, of purpose, and, of intelligence—one of an extraordinarily advanced nature.

A similar phenomenon applies to the origin of the first cell on Earth. It was once widely assumed that life could arise easily through simple chemical reactions in a primordial ocean, under an early atmosphere composed of inert gases. However, current scientific understanding reveals that even the most basic living cell must have been

---

[8]Ockham's Razor (sometimes spelled Occam or Ockam) is a principle of economy, also known as the principle of parsimony (*lex parsimoniae*). It is a methodological and philosophical guideline attributed to the Franciscan friar, philosopher, and scholastic logician William of Ockham (1280–1349).

extraordinarily complex. It would have needed not only the capacity to interpret genetic instructions for reproduction but also the capability to initiate that process—an act requiring a level of functional decision-making far beyond what could be expected from random chemistry alone.

The decision to self-reproduce—and the knowledge of how to carry it out, which forms the foundation of Darwin's proposed tree of life—demands an immense presence of something non-material, something that cannot be explained by chemistry or physics alone. That "something" is information.

This information, along with the capacity for living organisms to make purposeful decisions, lies beyond the explanatory power of current science. The complex, encoded instructions found in DNA remain a profound mystery for the naturalistic worldview. While science has advanced to the point of acknowledging that such information could not have arisen by mere chance, it still cannot account for its origin.

DNA functions for the cell in the same way that software functions for a computer—it directs, instructs, and enables essential processes. But just as software requires a programmer, the intricate information encoded in DNA implies the presence of a mind behind it. Only intelligence can produce meaningful, functional information. And that intelligence—so advanced and present from the very beginning—is none other than the Creator described in the first chapter of Genesis in the Bible.

I now turn to a second, equally important question: Does this Creator—call Him whatever you wish; I call Him God—communicate with us? Or did He simply create us and then withdraw, leaving us to our own fate?

Like any loving parent, God has maintained ongoing and public communication with us in four distinct ways: through His creation, through our inner sense of feeling or conscience, through very special individuals known as prophets, and finally, through His Son, Jesus Christ. The latest two forms of communication are recorded in the Bible.

So, when asked whether God speaks to us, the answer is a categorical yes. God does communicate, and He does so through this extraordinary book—a book whose wisdom surpasses all human understanding. This divine wisdom is evident throughout the Scriptures, from the concise and majestic narrative of the universe's creation to statements containing knowledge that would only be confirmed by science centuries later.

Consider, for example, the Bible's references to:

- The sphericity of the Earth (Isaiah 40:22)
- The vastness and expansion of the universe (Psalm 104:2)
- The innumerable stars in the sky (Jeremiah 33:22)
- The fact that the Earth "floats" in space (Job 26:7)
- The water cycle (Ecclesiastes 1:7)
- The weight of air (Job 28:25)
- And the composition of white light (Job 38:24)

These insights, embedded in ancient texts, show a depth of understanding that predates modern scientific discovery—further affirming that the Bible is more than just a historical or religious document. It is a channel of divine wisdom and communication.

The biblical narratives also contain prophecies fulfilled with astonishing accuracy, offering compelling evidence of divine authorship. For example:

- The invasion of Greece by King Xerxes (who reigned from BC 485–464) is clearly described in Daniel 11:2–12.
- The rise and fall of Alexander the Great are foretold in Daniel 8:5–8, 21–22 and Daniel 11:3–4.
- The specific fate of King Nebuchadnezzar, including the manner of his downfall, is outlined in Daniel 4.
- The struggles of King Antiochus the Great in his efforts to maintain control over the Holy Land against Egyptian resistance are found in Daniel 8.

Even more compelling are the numerous prophecies concerning the coming of Jesus Christ—covering extraordinary details such as:

- Where He would be born
- Who His parents would be.
- Where and how He would live.
- The manner and location of His death.
- His miracles.
- The nature of His relationships—both allies and enemies.
- How He would be remembered.
- His resurrection.
- His legacy.

These are not vague or general predictions; they are specific, detailed, and historically verifiable. This is not coincidence or luck.

What sets the Bible apart is that no other sacred text from the major world religions contains this level of scientific insight and historically fulfilled prophecy. This fact alone strongly suggests that we are not only dealing with a unique book, but also that we are aligned with the true faith.

It is evident that the Bible is not an ordinary book. Despite being written over a span of 1,700 years by more than fifty different authors, living in different eras, on three continents, and using three distinct languages, it maintains a remarkable consistency and unity. There are no contradictions in the three central themes it addresses—the Church, salvation, and the Kingdom of Heaven—topics that are inherently complex and often highly contentious.

Such coherence across time, culture, language, and individual perspective points to a profound truth: behind the diverse human voices that penned the Scriptures, there was one ultimate Author—God Himself—guiding the prophets and writers to faithfully convey His message to humanity.

The final and most crucial question is: "Can we trust that communication?" The answer rests on the foundation of one extraordinary event—the resurrection of Jesus Christ. His resurrection is the ultimate confirmation that He was indeed the Messiah, the Son of God. As such, everything He said can be trusted as the very word of our Creator, our heavenly Father.

The resurrection was not a minor occurrence; it was an apotheosis event—immense, powerful, and awe-inspiring. It provides both evidence and full assurance that the Bible is truly the word of God. Jesus Himself affirmed the authority of the Scriptures of His time—what we now call the Old Testament—by quoting from them on more than sixty occasions. His apostles did the same, drawing upon those texts to teach, interpret, and confirm the unfolding revelation of God's plan.

Considering the resurrection and the endorsement of Scripture by both Christ and His apostles, we can confidently trust that the Bible is a reliable and divinely inspired communication from God to humanity.

The resurrection of Jesus is the cornerstone of Christianity. As the Apostle Paul boldly declared: "And if Christ has not been raised, then our preaching is useless, and so is your faith" (1 Corinthians 15:14). Without the resurrection, the Christian message collapses. It is the event that validates everything Jesus taught, claimed, and accomplished.

In this work, without relying solely on faith, I present a compelling body of scientific, historical, and logical evidence that supports the resurrection as a real, historical event. Approaching the topic like a forensic investigation, I compile a series of assessments and arguments—all of which are verifiable through independent sources—to demonstrate that the biblical account of the resurrection withstands scrutiny.

This collection of evidence is not abstract or inaccessible. On the contrary, it invites readers to examine the facts for themselves and come to a reasoned conclusion: that the resurrection of Jesus Christ truly occurred, and in doing so, it places an undeniable seal of authenticity on the Gospel message.

Everyone has faith—whether atheist or believer. The real question is not whether we have faith, but rather: in what or in whom is that faith placed?

Our faith is not blind. We possess an abundance of evidence—scientific, historical, philosophical, and experiential—that allows us to respond with confidence and conviction to the three central questions explored in this book.

Can we deny the existence of the wind simply because we cannot see it? Of course not, because its effects are undeniable—we feel its force, see its impact, and hear its presence. The same logic applies to the answers to these profound questions:

- Does God exist?
- Does He communicate with us?
- Can we trust that communication?

The manifestations of God's presence, His message, and the reliability of that message are clearly observable to those willing to examine the evidence. Like the wind, they are unseen but undeniably real.

Apologetics is the discipline of explaining and defending our beliefs through rational, evidence-based arguments. As such, the term "faith" will be intentionally set aside in the development of this work—not because it lacks importance, but because the goal here is to engage the intellect using reason and verifiable information.

This undertaking is not simple. It involves addressing some of the most profound and challenging aspects of our worldview—topics that have stirred philosophical and theological debate for centuries. Yet the mission of apologetics remains clear: to present truths that can be grasped by reason, while also resonating with the deep intuitions of the human heart.

At the end of this book, the reader will find a set of appendices that I believe will help deepen their understanding of the topics discussed throughout the main chapters.

- **Appendix A** explores the question: Who is—and who is not—God? Here, I address common inquiries such as: Who created God? What is He like physically? What are His character and intellect like? What are His works and His legacy? I also tackle many of the recurring questions that often arise during one's Christian journey, providing thoughtful, reasoned answers aimed at both new and seasoned believers.
- **Appendix B** deals with numerical notation and probability. My experience giving lectures has shown me that when large numbers

are mentioned, their significance can vary widely in interpretation from person to person. The same applies to probabilities. Most people understand, for example, that winning a lottery is unlikely—but how unlikely is it? In many cases, even science has historically underestimated the odds due to a lack of information that is now available. When we apply this updated knowledge in mathematical terms, we are often confronted with a reality that sharply contrasts with what we were taught as children.

- **Appendix C** presents a brief overview of the "great story"—a narrative that traces the key events from the origin of the universe to the emergence of humanity. It is a concise but meaningful summary of the pivotal chain of events that made our existence possible, placing our story within the broader context of cosmic and biological history.

- **Appendix D** offers a summary of what I consider to be one of the most consequential legal cases in the history of the U.S. education system: Kitzmiller v. Dover Area School District (2005). This landmark trial resulted in a ruling that prohibited public schools in the United States from teaching any alternative to Darwin's theory regarding the origin of life and the diversity of species. I include it here because of its enduring impact on science education and the broader discussion around faith, science, and academic freedom.

It is now up to you, dear reader, to determine whether the arguments presented in this book enable a true reconciliation between the biblical narrative and the scientific discoveries of the past century. With the evidence laid out before you, you are invited to decide whether the Bible is simply an ancient collection of moral teachings and stories, valuable but human in origin—or whether it is something far greater: the divinely inspired word of our Creator, imparted to chosen individuals across generations to establish a bridge of communication between God and humanity.

The resurrection of Jesus Christ stands as the pivotal event. If it truly occurred—and the evidence strongly suggests that it did—it affirms that He was indeed the Son of God, and in doing so, validates the authority of all Scripture.

The choice is yours: to view the Bible as inspirational literature—or to receive it as the living message of a God who exists, speaks, and seeks relationship with you.

I hope this book awakens in you the desire and courage to ask questions, and to pursue their answers through thoughtful, serious inquiry into those subjects we often hesitate to explore—questions we've long carried in our hearts but perhaps never dared to voice, especially when it comes to our faith and religion.

Let us take inspiration from Mary, who, when the angel Gabriel announced, "Behold, you will conceive in your womb and bear a son," responded not with silence or blind acceptance, but with a humble and sincere question: "How will this happen?" Her example reminds us that faith is not incompatible with inquiry—that genuine understanding often begins with the courage to ask.

May we, too, give voice to our deepest questions—free from fear or misplaced apprehension—and seek answers through reflection, reason, and reverence.

DOES GOD EXIST?

For all men were inherently foolish who remained in ignorance of God, and did not come to know him who is, even while observing the good things around them, nor recognize the artisan while studying his works. To their way of thinking, either fire or wind or the swift air, or the periphery of the stars, or tempestuous water, or the luminaries of heaven were the gods that govern the world. If they have been deluded by the beauty of these things into believing that these were gods, let them come to understand how far superior to these is their Lord, since He was the source of beauty that fashioned them. And if they were astonished at their power and energy, let them realize from observing these things how much more powerful is He who made them. For from the grandeur and the beauty of created things is derived a corresponding perception of the Creator.

Wisdom 13:1-5

For centuries, humanity believed the Earth was flat, despite mounting evidence to the contrary. As early as the fourth century BC, Aristotle was among the first to describe the Earth as a sphere and even attempted to calculate its circumference. A few centuries later, the Greek mathematician Eratosthenes refined that calculation and was the first to suggest that the planet was tilted on its axis. By the 13th century, the most influential astronomical text of the time, *De Sphaera Mundi* by the Irish scholar Johannes de Sacrobosco—required reading at universities across the Western world—explicitly described the Earth as a sphere.

Yet, some individuals—skeptical of established knowledge—persisted in denying the evidence offered by astronomy, mathematics, and geography. Their eyes perceived a flat world, and they dismissed any suggestion to the contrary as absurd.

In the 20th century, a German intercontinental missile captured the first photograph of Earth from space on October 24, 1946, confirming beyond a shadow of a doubt what science had demonstrated for centuries: the Earth is indeed round.

Despite the overwhelming evidence, it is still surprising that some people today continue to deny the roundness of our planet. One of the most prominent figures among them was Samuel Shenton, a British citizen and member of both the Royal Astronomical Society[9] and the Royal Geographical Society[10]. In 1956, Shenton founded the Flat Earth Society[11], serving as its president and primary spokesperson until his death in 1971.

During his leadership, he made hundreds of appearances on television, at universities, and in conferences, passionately defending his beliefs. He attempted to refute all opposing evidence—including photographs of Earth from space—which he dismissed as either optical distortions caused by curved camera lenses or, in more extreme cases, deliberate fabrications.

Remarkably, the society he founded continues to exist and today boasts thousands of members, including some university professors and academics from various fields[12].

Just as some people refuse to acknowledge the roundness of the Earth despite clear and abundant evidence, others deny the existence of God—or a Creator—even in the face of compelling information. Yet, with the tremendous scientific advancements of recent years, it is becoming increasingly difficult to dismiss the notion that the universe is the product of intentional design.

---

[9]The Royal Astronomical Society was originally founded as the London Astronomical Society in 1820 to promote astronomical research. In 1831, it was granted the title "Royal" by King William IV and became known as the Royal Astronomical Society.

[10]The Royal Geographical Society is a British institution founded in 1830 under the name Geographical Society of London, with the aim of advancing geographic science. It was established under the patronage of King William IV of England.

[11]https://www.tfes.org

[12]Those interested in learning more about the Society's members can watch director Daniel J. Clark's acclaimed documentary Behind the Curve, available on Netflix.

Albert Einstein once remarked, "Man finds God behind every door that science manages to open."

Though Einstein was not a traditional believer, his use of the term "God" referred to a creative force—an intelligent origin behind all that exists. His insight stands in sharp contrast to the atheistic claim that everything we see is the result of random chance operating within nature. This idea contradicts the claim of atheists that the simple force of chance in nature is the cause of everything that exists.

I would like to clarify something about atheists. Just as there are professional soccer players who dedicate their lives to mastering the sport, there are also those who casually kick a ball around with their children in the backyard—yet believe themselves to be experts. In much the same way, atheism comes in many forms.

There are intellectual atheists, who rigorously examine evidence to support their views and engage enthusiastically in debate and discussion. There are activist atheists, who seek to persuade others to adopt their perspective. There are also antitheists, who may believe that "something"—rather than "someone"—created the universe, but view religion as inherently ignorant and consider any person or institution associated with it to be regressive or even harmful to society.

Finally, there are those who call themselves atheists more out of apathy than conviction—individuals who show little interest in questions of existence or faith, often driven by cultural attitudes or anti-religious sentiments rather than thoughtful consideration. They are disengaged and uninterested in learning more.

Throughout this book, when I refer to atheists, I include this entire spectrum. While they differ in approach and attitude, all share one central conviction: the belief that there is no compelling evidence for the existence of God.

The common atheist—the type you are familiar with—will often say, "Prove to me that God exists," or, more emphatically, "Prove to me scientifically that God exists." The inclusion of the word "scientifically" is meant to confer credibility, as if any statement preceded by it must be unquestionably true.

You could just as easily respond with, "Prove to me that God does not exist." But doing so would simply turn the conversation into a futile exchange —a game of fools, each believing they are clever, yet neither arriving at truth. Both participants remain equally ignorant, trapped in a debate with no productive outcome.

So, what exactly does someone expect when they ask for scientific proof of God's existence? Before attempting to answer that question, it is important to first reflect on what "scientific proof" means.

Most people readily accept the claim that aspirin is scientifically proven to relieve headaches. But how was this conclusion reached? Fundamentally, through statistical analysis. A series of controlled tests is conducted to observe and measure the outcomes of interest. These results are then analyzed to draw conclusions based on patterns and probabilities.

In the case of aspirin, researchers select a large and diverse group of individuals—varying in age, sex, race, and other factors. When these participants report having a headache, some are given aspirin while others receive a placebo[13]. The results are then documented: those who experienced relief, who did not, and who felt worse.

Suppose the results show that 70% of those who took aspirin felt better, 20% experienced no change, and 10% worsened. Meanwhile, in the placebo group, only 15% improved, 65% felt no better, and the rest deteriorated. What conclusion can we draw? Did the medicine work?

This kind of research forms the basis for the widespread belief that aspirin is effective. Yet, in the strictest sense, what the study reveals is that there is a high probability that aspirin relieves pain—not an absolute guarantee.

Even so, the term "scientifically proven" carries weight. It lends credibility and authority to the claim. When we hear, for example, that smoking causes cancer, we understand that this statement is backed by scientific evidence—specifically, long-term statistical and probabilistic studies. The accumulated data does not prove the connection in every

---

[13]For example, a sugar pill designed to look identical to the drug being tested.

individual case, but it overwhelmingly supports the conclusion that a causal relationship exists.

Returning to the subject at hand: when someone asks for proof of the existence of a being we cannot see—such as Cleopatra or God—what they are really asking for is evidence that strongly suggests such a being exists or once existed.

That is precisely what I aim to present in this chapter: a series of convergent and convincing scientific arguments that point to the manifestation of God. To do so, I will draw upon discoveries from astronomy, physics, and microbiology.

I have prayed to the Holy Spirit for discernment, asking for help in explaining these complex ideas in the simplest possible terms. Still, I recognize that some arguments may seem challenging or unfamiliar. You may even feel tempted to set the book aside. But I urge you: keep reading.

Even if you don't fully grasp every technical detail, I am confident that you will understand the central message—and that it will bring you joy to discover that there is, indeed, a reasoned and meaningful way to support what many have always intuitively believed: that God exists, at the very least, in His role as Creator.

And if, despite your best efforts, a particular argument feels unclear, simply move on to the next thesis. Each one stands on its own and contributes to the overall conclusion of this question.

In the remainder of the book, I shift the focus away from science and turn instead to the Bible—approaching it from a perspective that may be different from what you are accustomed to. If scientific topics are not your strongest interest, I trust you will find the following chapters much more familiar and spiritually enriching.

ARGUMENT: THERE IS NO DESIGN WITHOUT A DESIGNER.

Imagine walking along the beach of a deserted island when you come across a transparent container, apparently made of glass. It has a cylindrical body and a long, narrow neck. You instantly recognize it as a bottle, clearly created by human hands—not as a random product of the sea acting upon a piece of silicon.

Why do you make that judgment so quickly? Because your experience tells you that only humans—not the chaotic motion of waves—can produce such a plainly purposeful object. The shape, structure, and material all point to intelligent intervention, even though the object itself is simple.

Now let us leap forward several centuries into the future. Picture someone hiking through a dense forest in what is now Keystone, Pennington County, South Dakota. As this explorer moves through the trees, they stumble upon a granite mountain bearing the unmistakable faces of four former U.S. presidents. They have arrived at what we know today as Mount Rushmore.

Would this person believe that those facial features were carved into the mountain by natural erosion over time? Of course not. While erosion is known to alter landscapes significantly, it cannot sculpt recognizable human faces with that level of precision. The explorer understands, instinctively, that such forms must have been crafted by the hands of a sculptor—by intelligent design.

But let us push the thought experiment further. What are the mathematical odds that erosion could form such features, given millions of years? Would the probability be zero? Not quite. It might be something like one in a million trillion trillion trillion trillion..., continuing into near-infinity—but technically, not zero.

Still, does that infinitesimal probability give us a reason to doubt what is obvious? Do we seriously consider that erosion might have created Mount Rushmore?

Of course not.

Because the evidence of design is so clear, so unmistakable, and so compelling, we naturally and confidently conclude that intelligence—not chance—was behind it.

When Peruvian pilots first flew over the now-famous Nazca Lines[14] in the mid-20th century, their immediate conclusion was that the vast geoglyphs had been created by an ancient civilization. Who made them? How were they constructed? When were they created? These questions remained unanswered at the time—but one thing was certain: they did not believe these shapes were the result of natural forces acting on the desert floor.

Why were they so sure? Because experience had taught them that only intelligence and creativity—human intelligence—could produce such precise and deliberate forms. No one seriously entertained the idea that wind, erosion, or climate alone could have etched massive depictions of animals and geometric patterns across miles of land.

It is the same instinctive conclusion drawn by the hypothetical explorer at Mount Rushmore. Just as no one would attribute those presidential faces to the forces of erosion, no one who sees the Nazca Lines from above would think they were formed by chance.

---

[14]The Nazca Lines are a series of large-scale designs etched into the earth's surface, spanning a vast area of the Nazca Desert in the Ica region of Peru. They consist of approximately three hundred geoglyphs—ground drawings in the form of geometric, anthropomorphic, zoomorphic, and phytomorphic figures—ranging in length from fifty to three hundred meters.

These geoglyphs cover an area of about 450 square kilometers. The lines vary in width from fifteen to eighty inches, and their depth never exceeds twelve inches. They were created by removing the top layer of reddish, oxidized pebbles, revealing the lighter-colored soil beneath, which forms the visible designs when viewed from above.

Remarkably, the Nazca Lines have remained almost perfectly preserved over the centuries, largely due to the extremely dry climate of the region, which experiences minimal rainfall.

Once again, the design is obvious, the intent unmistakable, and the evidence of planning and purpose undeniable. The only reasonable explanation is that these figures were the product of intelligence.

William Paley[15], author of *Natural Theology* (published in 1802), is known for presenting the famous "watchmaker analogy." In his book, Paley argues that if one were to find an abandoned watch, the intricate arrangement of its parts would naturally lead to the conclusion that the components were intentionally designed and assembled for a specific purpose. In this case, the designer would be a watchmaker.

Paley then draws a parallel between the complexity of a watch and that of biological organs—most notably, the human eye. Just as a watch implies a watchmaker, the eye, with all its delicate structures and precision, implies the existence of a designer. His central point is simple yet powerful: complexity with purpose points to intelligent design.

Of course, the knowledge of astronomy and biology available in Paley's time was limited—so limited, in fact, that a modern elementary school student knows more about these subjects today than the greatest scholars of his era. Still, even though the how and when of creation were unknown in Paley's day, the evidence for design was so compelling that it affirmed, rather than weakened, belief in a Creator.

The logic remains just as forceful today:

- If there is a watch, there must be a watchmaker.
- If there is a building, there must be an architect.
- If there is a sculpture, there must be a sculptor.
- And if there is design, there must be a designer.

---

[15]A British philosopher and theologian who lived from 1743 to 1805.

Nature—which encompasses life, the universe, matter, and everything in between—follows a design. And where there is design, there must be a designer. For me, that designer is God.

You may choose to call this designer by a different name for now, especially if you find it difficult to associate the term "God" with Christianity or any religion. That is understandable. What is most important at this stage is to acknowledge the presence of a superior intelligence—a mind that established the laws of nature and embedded within them the information necessary to bring about life and shape the universe as we know it.

And just as every created thing reflects intention, this creation—which includes you and me—also has a purpose. Like every design, it was made for something greater than chance.

FIRST THESIS: A DIGITAL FACTORY INSIDE THE CELL

Understanding how life originated—and how the transition from non-living matter to living organisms occurred—has long remained one of humanity's greatest mysteries. For centuries, the only explanations available were rooted in religious belief. In the Western tradition, this explanation is found in the Bible, specifically in its first book, Genesis, which declares that God created all things.

Those who hold to this view—often referred to as creationists—do not claim that Genesis is a scientific account. Rather, we regard it as a divine revelation of God's creative work, forming the foundation from which we interpret the origins of all that exists, both visible and invisible. While Genesis is not framed as a scientific treatise, we believe there is no contradiction between its message and the discoveries of science to date.

Since the earliest days of Christianity, the first book of the Bible—Genesis—has inspired a wide range of interpretations, leading to diverse perspectives on its meaning. Some readers embrace a strictly literal view, understanding the "days" of creation as 24-hour periods. According to this interpretation, the entire journey from nothingness to an immense and complex material universe, including humanity, unfolded in just seven days. From that point on, the narratives of the Old Testament take shape. Those who hold this view believe the Earth is only a few thousand years old and are commonly referred to as young Earth creationists.

Others interpret the biblical "days" not as literal 24-hour intervals, but as symbolic periods—each spanning millions of years—during which God actively guided the development of matter and life. This perspective, which I share, is known as Old Earth Creationism. By adopting this interpretation, I have found it easier to reconcile Genesis' version with the findings of modern science. It allows for a view of the universe's development that is both divinely guided and consistent with empirical evidence.

Regardless of how one interprets the details of Genesis, most agree on the foundational truth proclaimed in its opening verse: "In the beginning, God created the heavens and the earth." This statement affirms that the universe had a deliberate and divine origin. In contrast, atheistic explanations remain unable to account for the origin of the universe's raw material—where it came from, and why it exists. As believers, we affirm that everything was created by God. He endowed matter with the information and order necessary for it to organize itself, giving rise to the cosmos and to life itself.

In the 18th century, most biologists accepted the Genesis account of creation. To them, the complexity and purposeful design evident in each species—what they termed "adaptation"—served as compelling evidence of God's intellectual authorship. Every organism seemed tailored to its environment in a way that revealed the signature of a divine Creator. Fish were equipped with gills for life underwater; birds had wings for flight; giraffes bore long necks to reach the highest leaves. Each species, it was believed, had been purposefully designed to thrive in its unique ecological niche.

This perspective changed dramatically following the publication of *On The Origin of Species* by British naturalist Charles Darwin on November 24, 1859. In this landmark work, Darwin proposed a revolutionary explanation for the origin and diversity of life on Earth. According to his theory, life began with a simple, primitive organism. As these early life forms reproduced, random errors—now understood as genetic mutations—occasionally occurred. For instance, an animal that climbed trees might be born with a slightly longer tail. This accidental variation provided a functional advantage, offering better balance or grip while evading predators. As a result, individuals with longer tails were more likely to survive and reproduce.

Over generations, these advantageous traits were passed down, gradually becoming more common within the population. As new mutations arose and accumulated over vast stretches of time, the cumulative changes could eventually become so pronounced that they gave rise to entirely new species. Darwin called this gradual, unguided process natural selection—the mechanism by which favorable traits become more prevalent in a population, leading to the evolution of species.

In Darwin's view, it was not divine intervention, but rather the blind force of nature that guided this process. Evolution through natural selection, he argued, could account for the remarkable diversity and complexity of life without invoking a Creator to explain the intricate "design" seen in nature.

One of the unintended consequences of Darwin's theory was its effect on humanity's perceived place in creation. According to the biblical account, man holds a uniquely privileged position in the universe. At the end of each stage of creation, Scripture tells us that "God saw that it was good" (Genesis 1:12, 18, 21, 25, 30). But when it came to creating humankind, God did something entirely distinct—He formed us in His own image and likeness (Genesis 1:27). What a profound and beautiful truth!

Darwin's theory, however, undermines this divine distinction. In his view, humans are not the pinnacle of creation but merely another product of random mutations and chance events over vast periods of time. Our form, intellect, and abilities are not the result of divine intention, but of evolutionary happenstance—just as the guava tree, by a different series of fortunate "accidents," developed guavas. We gained intelligence; the tree produced fruit. In this framework, the difference between us is simply the result of nature's lottery.

Darwin built part of his argument by drawing parallels between nature and the artificial selection practiced by farmers. In his day, animal breeders knew how to enhance desirable traits in livestock. For example, a sheep farmer who wanted woollier sheep would selectively breed the wooliest males and females. Over successive generations, the offspring would exhibit increasingly dense wool. In this process, human intelligence played a decisive role—it was the farmer who chose which animals would reproduce based on his goals.

Darwin proposed that nature could achieve similar outcomes without the involvement of any guiding intelligence. He used the example of a harsh winter to illustrate this idea: if only the wooliest sheep survived the cold, then they alone would be left to reproduce. If several such winters occurred, the surviving population would gradually become woollier—just as they would be under the farmer's guidance. The key difference, Darwin emphasized, was that no intellect was required. Nature itself, through selective pressure, could shape species over time.

This concept, which he called natural selection, replaced intentional design with a mechanistic, unguided process.

But Darwin went even further. He proposed that if additional major environmental changes occurred in the region where these sheep lived, the wooliest among them would continue to adapt in response. Over many generations, the cumulative changes could become so substantial that these sheep would no longer resemble their distant ancestors and would eventually be classified as an entirely new species. According to Darwin's theory, this branching process of gradual transformation could continue indefinitely, with each species giving rise to new ones over time. In this way, he claimed to offer a comprehensive explanation for the origin of all species on Earth.

However, Darwin's theory left two significant questions unanswered—questions that remain subjects of discussion even today:

- How is biological information transmitted from parents to offspring?
- If evolution favors progress, why have 99% of all species gone extinct?

These unresolved questions highlight some of the limitations of Darwin's original formulation and continue to provoke reflection in both scientific and philosophical circles.

To continue this discussion, it is important to clarify two key terms that often appear in conversations about evolution: microevolution and macroevolution.

Microevolution refers to small-scale evolutionary changes that occur *within* a species. These changes can often be observed over relatively short periods—sometimes within just a few generations. Microevolution includes variations in traits such as size, color, or physiological features, driven by mechanisms like mutation, natural selection, and genetic drift. Two frequently cited and widely accepted examples of microevolution include the *selective breeding of woolly sheep*, as discussed earlier, and *Darwin's observations of finches* on the Galápagos Islands, where variations in beak shape and size were seen as adaptations to different food sources. These changes occur *within the species boundary* and are empirically demonstrable.

Macroevolution, on the other hand, refers to large-scale evolutionary changes that go *beyond* the species level, resulting in the emergence of entirely new species, genera, or higher taxonomic groups over long geological timescales. Unlike microevolution, macroevolutionary processes are not directly observable within a human lifetime and are typically inferred from fossil evidence, comparative anatomy, and genetic analysis. A classic example cited by Darwin—and still referenced today—is the supposed evolutionary relationship between the modern whale and its closest living relative, the hippopotamus. According to this view, both species are believed to have evolved from a common ancestor, a small, four-legged, tailed, weasel-like mammal known as *Indohyus*, which lived approximately fifty-five million years ago[16].

It is important to note that while *microevolution* is widely documented and uncontested, *macroevolution* remains a topic of significant debate, particularly when it comes to the mechanisms and evidence supporting such vast transitions. Moreover, when most people refer to "the theory of evolution," they are often thinking specifically of macroevolution—the sweeping, long-term transformation of life from one form to another.

I will return to the topic of macroevolution later for a more in-depth exploration.

It is important to recognize that Charles Darwin never set out to explain the origin of the first form of life—the foundational spark from which all other life supposedly emerged. His focus was on addressing a different question: Why is there such an immense variety of life on Earth, in all its forms, scales, and adaptations? Darwin's goal in *On The Origin of Species* was to explain the diversity of life, not its initial emergence.

But why did Darwin omit the critical question of life's beginning— the origin of that first living cell? The answer lies, in part, in the

---

[16]The oldest whale skeletons discovered date back fifty million years and were found in what is now Pakistan. In contrast, the earliest known hippopotamus fossils are about fifteen million years old and were unearthed in southern Africa.

limitations of scientific knowledge and technology in the 19th century. At the time, biologists used optical microscopes capable of magnifying specimens up to around 2,000 times. While this was impressive for its time, it pales in comparison to modern electron microscopes, which can magnify objects up to ten million times, revealing astonishing details at the molecular and even atomic level.

Under their comparatively primitive instruments, 19th-century scientists observed cells as simple blobs of jellylike material encased in thin membranes. They referred to this substance as protoplasm and distinguished it from the cell's nucleus, which appeared denser and more defined. But when it came to the composition of protoplasm, they had little idea. To them, it was an amorphous chemical "jelly"—a shapeless mix without visible structure or discernible functionality. As a result, the complexity of the first cell seemed minimal, and its spontaneous formation appeared plausible.

From this perspective, it was easy for early biologists to speculate that Earth's primitive conditions—rich with various chemical compounds—could have naturally given rise to life. They imagined a "primordial soup[17]" filled with basic substances that, under the influence of lightning, radiation, or other environmental factors, combined by chance in just the right proportions. Out of this random assembly, they believed, a rudimentary cell eventually formed—one that somehow knew how to absorb nutrients, survive, and reproduce. From there, time and natural processes would do the rest, shaping that simple organism into the breathtaking diversity of life we see today.

This explanation, while imaginative, rested on the assumption that the first living cell was fundamentally simple. As we now know that assumption was deeply mistaken—an issue I will address in more depth later.

The theory of evolution is so deeply ingrained in people's minds that they refuse to accept any revision or clarification. Its dismissal of new

---

[17]"Primal broth," also referred to as "primordial broth," "primitive broth," "primordial soup," "prebiotic soup," or "nutritive broth," among other names, is a metaphor used to illustrate a hypothesis about the origin of life on Earth.

evidence uncovered by paleontologists and biologists in recent years is often automatic.

> The extreme rarity of transitional forms in the fossil record persists as the trade secret of paleontology. *The evolutionary trees that adorn our textbooks have data only at the tips and nodes of their branches; the rest is inference, however reasonable, not the evidence of fossils*[18]. (*Emphasis* mine).

Although the fossil record has not confirmed Darwin's theory of evolution—specifically macroevolution—since transitional fossils between distinct species have yet to be found, I will set this serious issue aside for now and proceed under the assumption that the theory is correct. I will accept the hypothesis that, through small and successive mutations, one species can gradually transform into an entirely different one. Even so, an extremely important problem remains unresolved.

The entire theory of evolution—specifically macroevolution—rests on the premise that cells occasionally make mistakes, or mutations, during the process of reproduction. If a mutation enhances an organism's ability to survive in its environment, natural selection ensures that this trait is passed on to the next generation. But what does it mean for a cell to reproduce?

Reproduction implies that the cell contains all the information required to copy every part of itself and assemble a complete, duplicate function. This intricate process results in two identical cells, each capable of continuing the cycle. Given this, a fundamental question arises: How did the very first cell—presumed to have formed by chance—acquire the complex information needed to reproduce in the first place? And further, how did it "decide" to initiate that process?

These are not trivial questions. In fact, they present a profound challenge for materialists and atheists, who must account for the spontaneous emergence of such extraordinary organization and purpose

---

[18]The quote is taken from the book Natural History by Stephen Jay Gould (1941–2002). Gould was a renowned American paleontologist, evolutionary biologist, and historian of science. He was also one of the most influential and widely read popular science writers of his generation. Throughout his career, he taught at Harvard University and worked at the American Museum of Natural History in New York.

from purely random, undirected processes. For us believers, however, this is not a mystery. We understand that such order and intent are not the products of chance, but of a Creator who imbued life with purpose and design.

From a naturalistic point of view, the origin of the first cell—with the necessary information to "know" how to reproduce and the apparent ability to "decide" to do so—is akin to a paradox. It resembles the story of a man walking through a deserted forest who accidentally falls into a fifty-meter-deep well. After much thought, he comes up with a solution: "Easy! I have a fifty-meter ladder at home. All I have to do is bring the ladder here and climb out." The flaw in this reasoning is obvious—to bring the ladder; he must first get out of the well, which is exactly the problem he is trying to solve.

This analogy reflects the dilemma faced by evolutionary theory. The theory may function after a cell already possesses complex information that enables it to reproduce, stay alive, and make coordinated "decisions." But how did that information arise in the first place?

By the time a cell has acquired such information, it can already perform a multitude of essential and highly organized functions. This reality is especially clear to us in the digital age. Consider how, once a smartphone has an operating system—whether Android, Windows, or iOS—and certain applications installed, it can carry out a wide range of tasks. Similarly, the cell requires its own "operating system" and numerous biochemical "applications" to function: it must find and process nutrients, maintain internal stability, self-repair, communicate with other cells, and most importantly, reproduce.

The crucial question, then, remains: Where did that original, vast repository of information come from? The idea that such complexity and purposeful organization could emerge by chance from random chemical interactions presents a serious problem for the naturalistic worldview. For believers, however, the existence of this information points clearly to an intelligent Creator—the source of both the design and the life that flows from it.

By applying the same scientific method that Darwin used to attempt a reconstruction of the distant past[19], we are justified in asking: What known source can produce the kind of complex, functional information required for that first living cell? When we examine all the sources known to humanity, there is only one answer: intelligence.

Until the middle of the last century, biologists understood that proteins—essential molecules in all living organisms—were composed of chains of amino acids, and that these chains played critical roles in maintaining life within the cell. During reproduction, for example, these amino acid chains convey information that helps the cell determine what type of tissue needs to be produced—whether skin (epidermis) or bone or any other kind of tissue.

When an injury occurs, nearby cells respond with remarkable coordination. Some produce protein with anticoagulant properties, which acts as a sealant upon contact with oxygen, helping to close the wound. At the same time, other cells begin producing the necessary components to regenerate tissue, such as new skin and muscle fibers. Through this intricate and highly regulated process, the body can repair damage and restore function—a testament to the complexity and precision of biological systems.

A single cell relies on thousands of different proteins to carry out its vast array of functions. For a time, biologists believed that what distinguished one protein from another—such as an anticoagulant protein versus a collagen protein—was simply the quantity of amino acids. They observed that the total number of amino acids in a protein varied depending on its function. For instance, yeast proteins typically contain around 466 amino acids, while titin—the protein responsible for tissue elasticity—has an astonishing 27,000 amino acids.

Based on this, many assumed that chance alone could account for the formation of these amino acid chains. If a protein had the correct

---

[19]It involved searching for the most logical cause—or source—to explain the existence of a given phenomenon. For example, if while digging, one discovered an extensive and deep layer of ash, the natural question would be: what known source could produce such a deposit? The most logical answer would be a volcano, as only a volcanic eruption is known to generate such a significant accumulation of ash.

count of each amino acid, it was believed that this would determine its type—forming, for example, an *X* protein instead of a *Y* protein.

However, this idea was entirely overturned in 1951, when Frederick Sanger[20] made a groundbreaking discovery. He demonstrated that proteins are not defined merely by the number of amino acids they contain, but by the precise sequence in which those amino acids are arranged. Functionality depends not just on complexity (the number of building blocks), but on specificity—the exact order of those components.

In other words, proteins are not just complex molecules; they are highly ordered and information-rich structures. The right amino acids must not only be present; they must be arranged in a very specific sequence to create a functional protein. This discovery added an entirely new dimension to the question of how such intricate biological information could have arisen by chance.

Let me illustrate this concept with a simple example. Imagine that instead of twenty (as in reality), there are twenty-seven amino acids available to form proteins[21]. For the sake of this analogy, we'll associate each amino acid with a different letter of the alphabet, including the space character.

Before the discovery of protein specificity, biologists believed that if a chain contained the correct number of each amino acid, it would result in a functional protein—regardless of their order. According to this view, the following two sequences of amino acids (represented here by letters) would have been considered the same protein:

- "es tsli bke ruahl cey"
- "ctu esey leh sb lrika"

---

[20]Frederick Sanger (Rendcomb, England, August 13, 1918 – Cambridge, England, November 19, 2013) was a British biochemist who was awarded the Nobel Prize in Chemistry twice.

[21]Proteinogenic, or natural, amino acids are those encoded by the genetic code. In most living organisms, there are twenty such amino acids: alanine, arginine, asparagine, aspartate, cysteine, phenylalanine, glycine, glutamate, glutamine, histidine, isoleucine, leucine, lysine, methionine, proline, serine, threonine, tryptophan, tyrosine, and valine.

Both contain the same number of each "amino acid": the same number of spaces, one "a," one "b," one "c," no "d," three "e", and so on. But then came Frederick Sanger's groundbreaking discovery, which revealed that the function of a protein depends not only on its composition, but on the precise sequence of amino acids.

To illustrate this, consider a third sequence:

- "the sky is clear blue"

This string contains the same number of letters and spaces as the previous two, but unlike them, it follows a specific, intelligible order that conveys a clear and meaningful message. In contrast, the first two are meaningless jumbles, despite containing the same characters.

This discovery was monumental. It showed that order matters—profoundly. Just as a coherent sentence in a written language requires the correct arrangement of letters, so does a functional protein require the exact arrangement of amino acids. Any deviation from this order destroys the function and renders the protein ineffective, just as random letters cannot convey a clear message.

This realization led biologists to a critical question: How does the cell "know" the exact order in which amino acids must be arranged to build a specific, functional protein?

This question cuts to the heart of modern biology—and raises profound implications about the origin of such intricate biological information.

As mentioned earlier, a functioning organism requires thousands of different types of proteins, each with a specific role. Even one of the smallest functional proteins is typically made up of about 150 amino acids. If there are only twenty types of amino acids available to form a protein[22], what is the probability that such a protein could arise by chance?

The total number of possible amino acid sequences for a protein of 150 amino acids is 20^{150}, which equals approximately 1×10^{195}. In other

---

[22]See two notes above.

words, the chance of a single, correctly ordered protein of this size forming randomly is one in 10^{195}—an astronomically small probability (see Appendix B).

Now consider this in the context of Earth's history. It is estimated that life appeared around 3.8 billion years ago, which is roughly 1.9×10^{17} seconds. If we divide the total number of permutations (10^{195}) by the number of seconds since life began, we get:

$$10^{195} \div 1.9 \times 10^{17} \approx 1 \times 10^{177}$$

This means that, for the correct protein to form purely by chance, the cell would need to:

1. Stay alive for 3.8 billion years, and

2. Attempt 10^{177} different amino acid combinations every second—just to form one functional protein.

And this scenario assumes success with only one protein of modest size. The cell requires thousands of proteins, many of which have 2,500 or more amino acids, making the probability of forming them by chance even more unimaginably remote.

The conclusion is clear: chance is not a viable explanation. The emergence and function of life requires far greater information. To evolve from a primitive cell to a human being does not involve the accumulation of random errors, but the acquisition and organization of vast, complex information.

So, the essential question remains: Where does the cell obtain this information?

The answer came just two years later, in 1953, when molecular biologists Francis Crick[23], James Watson[24], and Rosalind Franklin[25] uncovered the molecular structure of DNA[26] and its essential role in storing and transmitting information in living organisms. They discovered that DNA carries a genetic code—written in a precise sequence of chemical bases—that directs the synthesis of proteins.

The now-famous double helix structure, resembling a twisted ladder, contains the instructions necessary to build and maintain every protein required for life.

Each of the "steps" on the DNA ladder—called bases—represents a basic unit of information. These bases come in four types: adenine (A), guanine (G), thymine (T), and cytosine (C). Just like the zeros and ones in a computer's binary code, these four bases function as the fundamental elements of a biological information system.

In DNA, every three consecutive bases form a codon, which corresponds to a specific amino acid[27]. This means the sequence of bases along the DNA strand provides the instructions for assembling amino acid chains that fold into functional proteins. In this way, DNA serves as an

---

[23]Francis Harry Compton Crick (8 June 1916 – 28 July 2004) was a British physicist, molecular biologist, and neuroscientist. In 1962, he was awarded the Nobel Prize in Physiology or Medicine—alongside James Dewey Watson and Maurice Wilkins—for their discoveries concerning the molecular structure of deoxyribonucleic acid (DNA) and its significance in the transfer of information in living organisms.

[24]James Dewey Watson (born April 6, 1928, in Chicago) is an American biologist and co-recipient of the 1962 Nobel Prize in Physiology or Medicine.

[25]Rosalind Elsie Franklin (London, July 25, 1920 – London, April 16, 1958) was an English chemist and crystallographer. She contributed to the understanding of DNA's structure through X-ray diffraction images that revealed its double-helix form. She also made important contributions to the study of RNA, viruses, carbon, and graphite.

[26]DNA is the biomolecule that stores an organism's genetic information. It is a nucleic acid—specifically, deoxyribonucleic acid—composed of a sequence of nucleotides. Each nucleotide consists of a triphosphate group, a pentose sugar called deoxyribose, and one of four nitrogenous bases: adenine, cytosine, guanine, or thymine. The structure of DNA is a double helix, formed by two complementary and antiparallel strands.

[27]With a single base, only four amino acids could be specified; with two bases, sixteen combinations are possible. However, with three bases, up to sixty-four combinations can be formed.

instruction manual, containing all the "recipes" needed to produce each of the thousands of proteins an organism needs to function.

When the body suffers an injury and the cell needs to produce a coagulation protein, it "knows" exactly which segment of the DNA's three-billion-base sequence to access[28]. This specific segment, called a gene[29], contains the instructions for making that protein. The cell creates a copy of that gene in the form of RNA—a process known as transcription.

Once the RNA copy is made, it undergoes self-correction and processing, resulting in what is called messenger RNA[30] (mRNA). The mRNA then exits the nucleus and travels to the ribosomes, the cellular structures responsible for protein synthesis.

At the ribosome, the mRNA is "read" three bases at a time. Each triplet, or codon, corresponds to a specific amino acid. As the ribosome reads each codon, it selects the appropriate amino acid and links it to the growing chain. In this precise and highly coordinated process, the cell constructs the exact amino acid sequence needed to form the required coagulation protein.

Human DNA contains approximately three billion letters, representing an immense amount of information. To put that into perspective, it is equivalent to someone typing sixty words per minute, eight hours a day, for fifty years. Even the DNA of a simple, single-celled amoeba holds up to four hundred million bases of genetic information—enough to fill around eighty books, each five hundred pages long.

Bill Gates, the founder of Microsoft, once remarked, "DNA is like a computer program, but much, much more advanced than any software ever created." This comparison naturally raises a question: does a program create itself? Could all the applications on your phone—along

---

[28]Every cell in the human body—except for red blood cells—contains a DNA sequence that is approximately 3.2 billion base pairs long, equivalent to about two meters of DNA. To illustrate its density, a one-millimeter strand of DNA contains a sequence of over three million base pairs.

[29]The gene is the functional unit of heredity. Traditionally, it has been defined as a segment of DNA that contains the necessary information to produce a protein responsible for performing a specific function within the cell.

[30]RNA is another type of nucleic acid—specifically, ribonucleic acid.

with its operating system—come into existence simply by generating millions of random combinations of zeros and ones? Could something as sophisticated as *Facebook* or *WhatsApp* be the accidental result of random sequences?

The answer is clear. Just as software requires a programmer, the intricate information in DNA suggests the need for an intelligent source behind its origin.

While it is true that protobiologists[31] have proposed various hypotheses concerning the origin and development of the first cell—including one that suggests an extraterrestrial origin—they are grappling with a profound chemical mystery. Underlying the questions I have raised is a more profound enigma: how can matter become intrinsically purpose-driven? How can physical material be guided by coded instructions, a process that only makes sense within an intelligent framework?

Relying solely on protobiology to solve this puzzle is like trying to explain a book by detailing the chemical and physical processes involved in the production of paper and ink, while entirely ignoring the fact that the ink forms symbols, and that those symbols convey meaning to human beings. The medium is being studied while the message is being overlooked.

In 1954, Nobel laureate and prominent atheist George Wald[32] acknowledged this very tension when he wrote in *Scientific American*:

> The general opinion was to believe in spontaneous generation; the other alternative was to believe in supernatural creation. There is no third position.
> *Most modern biologists, having surveyed with satisfaction the fall of the spontaneous generation hypothesis, are still unwilling to accept the alternative belief of special creation, leaving themselves with nothing [...] When it comes to the*

---

[31]Protobiologists are scientists who study the earliest forms and origins of life, including primitive biological structures and processes.

[32]George Wald (New York, November 18, 1906 – Cambridge, Massachusetts, April 12, 1997) was an American scientist renowned for his research on retinal pigments. In 1967, he was awarded the Nobel Prize in Physiology or Medicine, along with Haldan Keffer Hartline and Ragnar Granit.

origin of life only there are two possibilities: creation or spontaneous generation. There is no third option.
Spontaneous generation was disproved a hundred years ago, but that only leads us to only one different conclusion: that of supernatural creation.
We cannot accept that for philosophical reasons; therefore, we choose to believe the impossible: that life arose spontaneously by chance! (Emphasis mine).

Four years later, in an article titled "Biology and Innovation," published in Scientific American, George Wald reiterated his argument. Despite acknowledging the profound challenges involved in explaining the origin of life through purely natural processes, he chose to reject what he referred to as the "only possible conclusion"—the existence of God. Instead, he embraced what he admitted was "scientifically impossible," driven not by evidence, but by a personal unwillingness to accept the idea of a Creator:

Spontaneous generation, [the idea that] life arose from nonliving matter, was scientifically refuted 120 years ago by Louis Pasteur and others. That leaves us with the only possible conclusion that life arose as a supernatural creative act of God.
I won't accept that philosophically, because I don't want to believe in God; therefore, I choose to believe what I know is scientifically impossible; spontaneous generation as something arisen from evolution. (*Emphasis* mine).

During the 1980s, in his article "Life and Mind in the Universe"—which he presented at the First World Congress for the Synthesis of Science and Religion in 1986, held in Bombay—George Wald made a notable shift in tone. In that piece, he acknowledged the limitations of materialistic explanations for the origin of life and began to entertain the idea that consciousness, or mind, might play a fundamental role in the universe. He wrote:

I have come to the end of my scientific life facing two great problems. Both are rooted in science; and I approach them as only a scientist would. Yet I believe to be in essence unassimilable as science. That is scarcely to be wondered at, since one involves cosmology, the other [the origin of] consciousness. The problem of consciousness was hardly avoidable for someone like me, who has spent most of his scientific life working on mechanisms of vision. That is by now a very active field, with thousands of workers. We have learned a lot and expect to learn much more; yet none of it touches or even points however tentatively in the direction of telling us what it means to see.

The retina of a frog is very much like a human retina. Both contain two kinds of light receptors, rods for vision in dim light and cones for bright light; the visual pigments are closely similar in chemistry and behavior; both have the same three fundamental nerve layers, and the nervous connections to the brain are much alike. But I know that I see. Does a frog see? It reacts to light —so does a photocell-activated garage door. But does it know it is responding, is it aware of visual images?

There is nothing whatever that I can do as a scientist to answer that question. [...] So that is the problem of mind — consciousness— a vast, unchartable domain that includes all science, yet that science cannot deal with, has no way of approaching.

The second problem is related to the special properties of our universe. [...] There is good reason to believe that we are in a universe permeated with life, in which life arises, given enough time, wherever the conditions exist that make it possible. [...] How did it happen that, with what seem to be so many other options, our universe came out just as it did?

A few years ago, it occurred to me —albeit with some shock to my scientific sensibilities— that my two problems, that of a life-breeding universe, and that of consciousness that can neither be identified nor located, might be brought together. That would be with the thought that mind, rather than being a late development in the evolution of organisms, had existed always: that this is a life-breeding universe because the constant presence of mind made it so. [...] Of course, implicit in such talk is the recognition that a universe in which mind can eventually achieve such overt expression as in science, art and technology must be at its core, from its inception, in some sense a knowing universe; that it must in some sense possess mind as its pervasive and enduring attribute. (Emphasis mine).

If an atheist were to arrive on what he believes is a deserted island and find the word "welcome" carved into the sand, he would have no reason to assume that the waves or natural forces produced the message by chance. He would immediately recognize that this sequence of characters was created by intelligence with the intention of communicating a message.

Now, a simple yet powerful question can be posed: if you acknowledge—without hesitation—that the only plausible source of that seven-character string of information is intelligence, why do you not apply the same reasoning to the DNA molecule? Unlike the brief message in the sand, DNA contains a coded sequence not seven characters long, but three billion. Why is the first example universally accepted as the product of intelligence, while the second, infinitely more complex, is attributed to chance and natural processes?

SECOND THESIS: MOLECULAR MACHINES

A single letter of the alphabet is specific, but not complex. Scatter thousands of letters randomly across a table, and you have complexity without specificity. A poem by Pablo Neruda[33], on the other hand, embodies both: it is richly complex and precisely specific. In much the same way, proteins are both complex and specific—they require a precise sequence of amino acids to function properly. Now, let us explore another captivating idea: irreducible complexity.

Michael Behe[34], the originator of the concept, defines an irreducible system as one composed of several well-coordinated parts that interact to perform a fundamental function. If even a single component is removed, the entire system ceases to operate. A classic example is the inner workings of a watch, which depend on the seamless interaction of pinions, cogwheels, gears, and springs. If any one of these elements is taken away, the mechanism fails.

The core idea behind irreducible complexity is that all essential parts must be present simultaneously. None of them can emerge gradually. For instance, it is not plausible to assume a pinion began with just two teeth

---

[33]Pablo Neruda, the pseudonym of Ricardo Eliécer Neftalí Reyes Basoalto (Parral, July 12, 1904 – Santiago de Chile, September 23, 1973), was a Chilean poet and recipient of the Nobel Prize in Literature in 1971. He is regarded as one of the most prominent and influential literary figures of the 20th century. Gabriel García Márquez once described him as "the greatest poet of the 20th century in any language."

[34]Michael J. Behe (born January 18, 1952, in Altoona, Pennsylvania) is an American biochemist known for advocating intelligent design. He is a professor of biochemistry at Lehigh University in Pennsylvania and a senior fellow at the Discovery Institute's Center for Science and Culture.

and "evolved" over time to reach the forty-eight teeth required for functionality—nor that the rest of the mechanism patiently awaited its final form. How would that pinion "know" it needed forty-eight teeth? If it did possess such knowledge, wouldn't that suggest a form of self-awareness or intentionality?

Such a "sense of purpose" cannot be generated by random, unguided processes. Rather, it implies direction from an external source. A clock can only function as such when all its internal components are simultaneously operational. There is no room for gradual assembly in its core mechanism. However, non-essential features—like the glass face or wristband—could develop gradually without affecting the clock's primary purpose: timekeeping.

An irreducible system consists of the minimum set of interdependent parts necessary to perform its function. Remove one, and the system fails entirely.

In biology, we also find compelling examples of irreducible systems—such as the bacterial flagellum, the visual system, the blood clotting cascade, the intracellular protein synthesis machinery, and the immune system among others.

If I were to show you a diagram of the bacterial flagellum without context and ask for your interpretation, you might conclude that it is a schematic of an electric motor from a ship. The resemblance is striking. Both systems feature a central shaft, elbow joints, rings, gears, a rotor, a stator[35], bushings, ball bearings, and a propeller—though in the case of the flagellum, the propeller takes the form of a whip-like tail.

The capabilities of this biological motor are nothing short of astounding. The flagellum can spin between 6,000 and 17,000 revolutions per minute, propelling the bacterium at speeds of up to sixty body lengths per second. To put this into perspective, the cheetah—the fastest land animal—reaches only about twenty-five body lengths per

---

[35]The stator is the stationary component of a rotating machine and one of the two essential elements for transmitting power in electric motors or generating electric current in generators. Its counterpart, the rotor, is the moving part that interacts with the stator to perform these functions.

second. Even more impressively, the flagellum can reverse its rotational direction in less than one one-hundred-thousandth of a second.

The motor's mechanical portion is constructed from at least twenty distinct proteins, while an additional thirty proteins are required to facilitate its function—pumping ions through rings and coordinating movement. All this intricate machinery is housed within a structure that measures just twenty millionths of a millimeter.

As with a ship's motor, the removal of any single component renders the entire system inoperable. Every part must be present and functional simultaneously for locomotion to occur. This raises a fundamental question: how could such a finely tuned system evolve gradually? How does a gear evolve piece by piece? Can we honestly say that chance alone is a sufficient explanation for the origin of a mechanism so precisely engineered?

Just as no one would attribute the creation of a ship's electric motor to random processes, it seems reasonable to ask whether the bacterial flagellum is not also the result of intentional design—an unmistakable sign of a purposeful designer.

Another compelling example of irreducible complexity is the blood coagulation system. When a blood vessel is injured, a multi-phase defense mechanism is rapidly initiated to prevent excessive blood loss. First, the vessel walls constrict to minimize blood flow to the affected area. Next, platelets—specialized blood cells—adhere to the site of injury and begin spreading across the vessel's inner surface. Simultaneously, tiny granules within the platelets release chemical signals that attract additional platelets, forming what is known as a platelet plug.

On the surface of these activated platelets, a series of complex biochemical reactions—collectively called the coagulation cascade—unfolds. This cascade involves a tightly regulated sequence of steps that culminate in the production of a fibrin clot. Fibrin acts like a biological net, stabilizing the plug and effectively sealing the wound to stop bleeding.

Seventeen distinct proteins are involved in this cascade, each one activated in a precise sequence to trigger the next step in the process.

Remarkably, these coagulation factors normally circulate in the blood in an inactive form, poised to act instantly when needed. The coordination required is extraordinary—comparable to a sophisticated computational system. It is as if an internal "computer" analyzes vast amounts of biological data to determine the exact moment and location for each protein's activation. All this information is coded in our DNA, which serves as the blueprint for this entire life-preserving mechanism.

This raises a profound question: How could such a system evolve gradually, step by step? How does it "know" the exact sequence of actions required to function flawlessly? Is it reasonable to attribute the origin of such a coordinated, life-critical process to random chance?

Or, like any well-orchestrated system, does it more plausibly point to purposeful design—a system crafted by an intelligent designer with a specific and vital function in mind?

Critics of irreducible complexity often argue that if a component—such as a pinion—were missing a tooth, the resulting device might still serve another function, such as acting as a paperweight. But this line of reasoning inadvertently reinforces the core argument of irreducible complexity: for a clock to function as a clock, each of its parts must be fully formed and properly integrated. A pinion with a missing tooth in a watch does not tell time; it merely occupies space. The moment the watch loses its ability to perform its intended function; it ceases to be a timepiece in any meaningful sense.

This leads to a deeper question: what would "motivate" a malformed or non-functional component to continue evolving toward a fully operational mechanism? How could an incomplete device—serving no relevant function—"know" what it is supposed to become? If such an object were to resume an evolutionary path toward functionality, would that not imply a kind of self-awareness or an inherent "sense of purpose"? These are attributes we typically associate with intention, not randomness.

The continued advancement of electron microscopy, now capable of magnifying up to ten million times, has opened an unprecedented window into the internal architecture of cells—both bacterial and human. What we see under these lenses is nothing short of astonishing:

elaborate, highly regulated systems working in perfect harmony, like miniature factories. Inside each cell, vast assembly lines operate with precision, guided by billions of instructions encoded in a molecular alphabet—DNA. It is a level of sophistication that would make any computer engineer envious.

Consider the construction of something as common as a cell phone. Its functionality depends on code—meticulously written, interpreted, and executed. No one would seriously propose that such a device could be the product of unguided chance. And yet, the inner workings of a single living cell far exceed the complexity of any manufactured technology.

Charles Darwin himself wrote:

> If it could be demonstrated that any complex organ existed, which could not possibly have been formed by numerous, successive, slight modifications, my theory would absolutely break down. But I can find no such case[36].

However, with the extraordinary insights provided by modern molecular biology, we have begun to uncover precisely such cases. Irreducibly complex systems—beyond the reach of gradual, step-by-step evolution—are not theoretical anymore. They are observable, real, and increasingly impossible to ignore.

THIRD THESIS: THE GREAT CAMBRIAN EXPLOSION

The Cambrian Explosion, as described by Stephen Jay Gould—renowned paleontologist, Harvard professor, and co-director of the American Museum of Natural History—stands as one of the most significant and mysterious events in the history of life on Earth.

To understand this phenomenon, we must first define the term Cambrian. Just as human development is categorized into stages—infancy, toddlerhood, preschool, and so on—Earth's history is divided into chronological segments known as eras, which are further subdivided into periods. The Cambrian period began approximately 550 million

---

[36]*On The Origin of Species*, Chapter VI.

years ago and marked a geological era that spanned roughly fifty-five million years. It is within this window of time that something extraordinary occurred: a sudden and unprecedented emergence of complex life forms.

There are locations around the world where the Earth's layered history is beautifully and visibly preserved. One of the most striking examples is the Grand Canyon in Colorado, where vividly colored and distinctly layered strata reveal a complete sequence of geologic periods. Another such site is northern Wales in the United Kingdom—a region that played a key role in paleontological history.

It was there that Charles Darwin, shortly after graduating from Cambridge University, first encountered fossils of complex organisms from the Cambrian period—creatures that already possessed nervous, digestive, circulatory, muscular, and reproductive systems. He was accompanied by his mentor, Professor Adam Sedgwick[37], a leading geologist of the time and one of the foremost experts on Cambrian fossils. Like many paleontologists of his era, Sedgwick was deeply familiar with the rich fossil record of the Cambrian, a record that challenged many conventional understandings of how life developed on Earth.

The sudden appearance of highly developed life forms in the fossil record—without apparent evolutionary precursors in earlier strata—continues to provoke scientific debate and fascination. The Cambrian Explosion remains a central puzzle in our understanding of life's origins.

Prior to the Cambrian period, no fossils of complex life forms had been discovered. All known fossils were from the Cambrian period onward. This absence puzzled the scientific community deeply—including Charles Darwin himself. His theory of evolution by natural selection predicted that every living species must have descended from simpler ancestral forms. So where were the fossilized ancestors of the many complex organisms that suddenly appeared in the Cambrian strata? Where were the remnants of evolutionary "experiments" that

---

[37]Adam Sedgwick (March 22, 1785, Dent – January 27, 1873, Cambridge) was a British geologist and one of the founders of modern geology. He is best known for his studies of the geological strata of the Devonian and Cambrian periods.

failed—those transitional forms that natural selection supposedly filtered out?

Darwin openly acknowledged this challenge in *On The Origin of Species*, writing:

> To the question why we do not find records of these vast primordial periods, I can give no satisfactory answer.

He recognized the Cambrian explosion as a serious difficulty for his theory. During this brief geological window, nearly 90% of all known animal families suddenly appeared in the fossil record—a biological event so abrupt and widespread that the term explosion is fitting.

To illustrate the timeline of life more vividly, imagine compressing Earth's 3.8-billion-year biological history into a single 24-hour day. In this analogy:

Life begins at midnight, with the appearance of the first unicellular organisms.

- At 6:00 a.m., only single-celled life exists.
- At 1:00 p.m., still no change.
- At 6:00 p.m., still only single cells inhabit the Earth.

For over 75% of the day, life consists solely of simple, unicellular organisms. Then, at 8:50 p.m., in the span of just two minutes, nearly all major animal body plans and integrated systems emerge—such as the nervous, circulatory, digestive, respiratory, reproductive, and skeletal systems, along with features like brains, eyes, and limbs. Even more remarkable is that these complex structures have remained largely unchanged for hundreds of millions of years.

The fossil record shows that during the Cambrian era, organisms appeared already equipped with the essential biological systems still present in animals today—including nervous, immune, excretory, lymphatic, endocrine, muscular, and others. Strikingly, no entirely new systems have developed since, nor are there fossils of creatures displaying incomplete or transitional versions before this period.

Furthermore, this pattern did not occur only once. In subsequent eras, the remaining 10% of animal groups also appear suddenly in the record, complete with the same nine systems. Again, no fossil traces of inferior or transitional forms have been discovered to precede these appearances.

This stark reality poses a significant challenge to Darwin's famous "tree of life," often featured in school biology textbooks. The fossil evidence does not show a gradual branching from a common trunk through a web of intermediate forms. Instead, what we find in the geological record resembles a collection of fully formed branches, abruptly emerging without visible connections to a shared trunk or root.

The concept of *variety* should not be confused with *macroevolution*. Over time, we have observed an increase in variation within species—for example, the emergence of new dog breeds adapted to different geographies and climates. This is an example of adaptation and microevolution, where traits shift within certain boundaries. However, the fossil record consistently points to a single, recognizable kind: the dog.

Though evolutionary theory posits that dogs evolved from more "primitive" ancestors, no transitional fossils conclusively documenting this transformation have been found. Instead, the structural integrity and form of the species remain consistent throughout time. Fossils attributed to dogs show abrupt appearances in the record, not gradual development from one kind into another.

These findings lend credence to the theory of creation, which maintains that organisms appear fully formed, according to distinct kinds, rather than evolving gradually from a common ancestor. Supporting this view is the discovery of fossils that date back tens or even hundreds of millions of years yet display minimal or no significant difference from their modern counterparts. In many cases, these ancient fossils are indistinguishable from the living species we see today.

Such evidence challenges the assumption that macroevolutionary changes have occurred over long periods and instead aligns more closely with the idea that species were introduced in complete form, with built-in capacity for variation, but not for transformation into entirely new kinds.

Darwin's contemporaries accepted the idea of microevolution—the notion that minor changes within a species could explain observable variation. For instance, Darwin's famous observations of the Galápagos finches, which exhibited notable differences primarily in the shapes and functions of their beaks, were widely seen as examples of adaptation to environmental conditions. These differences were understood to occur within a species, not to transform it into a fundamentally different kind.

What was revolutionary in Darwin's theory, however, was not the idea of variation, but the claim that all life originated from a single common ancestor. From this idea, Darwin developed his concept of the "tree of life," a branching diagram in which every organism traces its lineage back to a shared origin. The two central pillars of this theory— natural selection and common ancestry—have since become foundational to modern biology.

Despite the substantial and growing body of evidence that challenges key aspects of these concepts, such as the lack of transitional fossils and the abrupt appearance of complex organisms in the geological record, Darwin's framework continues to be widely taught and accepted. The "tree of life" remains a powerful symbol, even though fossil evidence often resembles more of a forest of disconnected trees—each kind appearing suddenly, fully formed, and without clear evolutionary precursors. Darwin documented in his book:

> There is another and allied difficulty, which is much more serious. *I allude to the manner in which species belonging to several of the main divisions of the animal kingdom suddenly appear in the lowest known fossiliferous rocks.* Most of the arguments have convinced me that all the existing species of the same group are descended from a single progenitor, apply with equal force to the earliest known species. For instance, it cannot be doubted that all the Cambrian and Silurian trilobites are descended from someone crustacean, which must have lived long before the Cambrian age, and which probably differed greatly from any known animal. Some of the most ancient animals, as the Nautilus, Lingula, etc., do not differ much from living species; and it cannot on our theory be supposed that

these old species were the progenitors of all the species belonging to the same groups which have subsequently appeared, for they are not in any degree intermediate in character.

Consequently, if the theory be true, it is indisputable that before the lowest Cambrian stratum was deposited, long periods elapsed, as long as, or probably far longer than, the whole interval from the Cambrian age to the present day; and that during these vast periods the world swarmed with living creatures.

Here we encounter a formidable objection; for it seems doubtful whether the earth, in a fit state for the habitation of living creatures, has lasted long enough.[38] (*emphasis* mine)

If we closely examine one of the Cambrian trilobites—as referenced in Darwin's own writings—we find that these ancient organisms possessed all the major biological systems previously mentioned: nervous, muscular, digestive, circulatory, reproductive, and more. This clearly indicates that by the time trilobites emerged, they already had highly complex DNA, containing millions of instructions essential to produce a vast array of specialized proteins. These proteins, in turn, orchestrated the formation of over fifty distinct tissue types, including a hard exoskeleton, compound eyes, a brain, muscular systems, stomach, antennae, and more.

The leap from a simple Precambrian bacterium—a single-celled organism—to a fully formed trilobite represents a staggering increase in biological complexity. Such a transition would require not only a massive expansion in genetic material but also an incredibly precise orchestration of developmental processes. This raises a fundamental question: Where did all this information come from?

Can this explosion of genetic and structural complexity be realistically attributed to random mutations and natural selection alone? Is it the product of extraordinary chance—an evolutionary jackpot? Or does the sheer volume and specificity of biological information require point more convincingly to the presence of design?

---

[38] On *The Origin of Species,* Chapter x.

FOURTH THESIS: THE FINELY TUNED UNIVERSE

Let us begin with something simple and familiar: a cake—a staple at celebrations and gatherings of all kinds. It may seem like an unlikely starting point, but bear with me; this everyday example will serve to illustrate a much deeper concept shortly.

A cake recipe might include:

- One billion quadrillion (1 followed by 32 zeros) particles of wheat flour.
- ½ cup of butter
- 1½ cups of refined sugar
- 1 cup of milk
- 3½ teaspoons of yeast
- 1 teaspoon of salt
- 1 teaspoon of vanilla extract
- 3 eggs

The preparation is straightforward:

- Preheat the oven to 180°C (350°F) and grease a 23 x 33 cm baking pan. Mix the salt and yeast into the flour and set aside.
- Cream the butter and sugar in a large bowl until fluffy. Add the eggs one by one, mixing well after each.
- Alternate adding the flour mixture and the milk, beating until smooth. Stir in the vanilla.
- Pour into the pan and bake for 45 minutes.

Now, imagine you could count every single flour particle. Here is the question: if you accidentally added one extra particle, or left one out, would that microscopic error ruin the entire cake? Would it collapse in the oven or become inedible?

Of course not. A cake is forgiving. Tiny deviations in quantity do not alter the result in any meaningful way.

And yet, as we will soon see, when it comes to the formation of the universe, the situation is very different. In that case, an error as small as a single figurative "flour particle" could have made all of existence impossible. The cake is just an analogy—but it prepares us to understand the astonishing precision required at the dawn of the cosmos.

Let us explore that in more detail.

When you hear the word "atom," you might immediately picture a cluster of spheres at the center, with smaller spheres orbiting around them in concentric circles. If so, you are visualizing the Bohr model of the atom, proposed in 1913 by Danish physicist Niels Bohr[39].

According to this model, the atom consists of a dense central nucleus made up of protons and neutrons, with electrons revolving around it in distinct energy levels or "shells." In this structure:

- Protons carry a positive charge.
- Electrons carry a negative charge.
- Neutrons are electrically neutral.

Bohr's model was groundbreaking at the time, offering a simple and intuitive representation of atomic structure. While later developments in quantum mechanics would refine and expand our understanding, this iconic image of orbiting electrons still shapes how many people imagine the atom today.

The concept of the atom—a fundamental, indivisible unit of matter—dates to ancient Greece, where it arose more from philosophical reasoning than empirical science. Thinkers like Democritus proposed that all matter was composed of tiny, indivisible particles called atoms, not based on experiments, but as a logical necessity to explain the nature of change and continuity in the physical world.

It was not until many centuries later, in the early 19th century, that the idea began to take on scientific form. In 1804, John Dalton proposed that all atoms of a given element are identical in mass and properties,

---

[39]Niels Bohr was awarded the Nobel Prize in Physics in 1922. He was born in Copenhagen, Denmark, in 1885 and died there in 1962. Bohr contributed to the Manhattan Project, participating in the development of the first atomic bomb in the United States. Throughout his career, he frequently engaged in intellectual debates with Albert Einstein, particularly on the interpretation of quantum mechanics.

and distinct from the atoms of any other element[40]. This marked a pivotal moment in atomic theory, initiating the systematic study of chemical behavior at the atomic level.

As knowledge progressed, scientists began cataloging elements based on their atomic properties. In 1869[41], Dmitri Mendeleev published the first organized inventory of elements arranged by atomic mass—the forerunner of today's periodic table. Mendeleev's table not only organized known elements but also predicted the existence and properties of undiscovered ones with remarkable accuracy.

The next major step in atomic theory was an attempt to visualize the structure of the atom. This gave rise to various atomic models, each attempting to explain observed chemical and physical phenomena. Among the most influential was Niels Bohr's model (1913), which depicted electrons orbiting a central nucleus in defined paths or shells, laying the foundation for quantum theory.

These evolving models represent a crucial shift from philosophy to science—transforming the atom from an abstract idea into a central pillar of modern chemistry and physics.

Each new scientific discovery about the atom opened the door to even more unanswered questions—many of which remain unresolved to this day. Among the most pressing mysteries that intrigued early physicists was a fundamental one: What gives the atom its stability?

Why do protons and neutrons, with their considerable mass, bind together in the nucleus instead of drifting apart? What force causes them to coalesce and stay compacted at the center of the atom? Similarly, how does the electron manage to orbit the nucleus indefinitely, neither spiraling inward toward the proton nor flying off into space?

This becomes even more perplexing when we consider electromagnetism, one of the most well-understood laws of physics.

---

[40]This was a postulate of the English chemist, physicist, and mathematician John Dalton (1766–1844).

[41]This was the work of Russian chemist Dmitri Ivanovich Mendeleev (1834–1907).

According to it, like charges repel, and opposite charges attract. If that is the case, then:

- How can multiple positively charged protons coexist in the nucleus without repelling each other violently?
- And if the electron carries a negative charge while the proton is positive, why don't they simply collapse into each other under the force of attraction?

These questions revealed that something beyond electromagnetism must be at work—something strong enough to counteract the repulsive forces within the nucleus and delicate enough to keep electrons in dynamic balance at a precise distance.

Physicists eventually proposed the existence of the strong nuclear force to explain this—but even this "solution" only raised more questions about the fine-tuned nature of physical constants and the underlying principles that govern matter. The more we uncover about atomic structure, the more we are confronted not just with complexity, but with a remarkable precision that seems anything but accidental.

In the 20th century, scientists discovered a force operating within the heart of the atom—the strong nuclear force[42]. This force is far more powerful[43] than electromagnetism, and it plays a vital role in atomic stability. Without it, protons, which all carry positive electric charges, would naturally repel one another and fly apart. But the strong nuclear force overcomes this repulsion, binding protons, and neutrons together within the nucleus and allowing atoms to exist.

But what if this force were altered—even slightly?

- If the strong nuclear force were eliminated, protons would no longer be held together. The nucleus would disintegrate, and atoms would cease to exist entirely. No atoms mean no matter— no stars, no planets, no life.

---

[42]This is one of the four fundamental forces acting between subatomic particles. The other three are the electromagnetic force, the weak nuclear force, and the gravitational force.

[43]The strong nuclear force is approximately 137 times stronger than the electromagnetic force acting between protons.

- If the force were slightly stronger, it would overpower the balance between the nucleus and the orbiting electrons. The electron could be pulled into the nucleus, merging with the protons and destroying the atom's structure. Again, atoms would not exist.
- Conversely, if the strong nuclear force were slightly weaker, it would no longer be able to hold the protons together. The electromagnetic force—which pushes like charges apart—would dominate, causing the nucleus to fly apart. Atoms would collapse before forming.

Without atoms, there can be no molecules. Without molecules, chemistry cannot occur. And without chemistry, there would be no stars, no planets, no life, and no universe as we know it.

In short, the strong nuclear force must have precisely the right value—not too strong, and not too weak—to ensure the stability of the atom. Its delicate balance is one of the most striking examples of fine-tuning in the universe. The existence of matter itself hinges on a force that is, quite literally, just right.

Gravity, also known as gravitation, is one of the four fundamental forces of nature. It is the force that draws two objects with mass toward one another. Though it is vastly weaker than the strong nuclear force, it plays an essential role in the formation and structure of the universe.

After the Big Bang, the universe was composed almost entirely of hydrogen atoms—the simplest atoms, each made of just one proton, neutron, and electron. Despite gravity's comparative weakness, it was just strong enough to begin pulling nearby hydrogen atoms toward one another. As atoms gathered, they formed clumps of matter. These clumps had more mass, which in turn generated more gravitational pull, allowing them to attract atoms that were farther away. Over millions of years, these growing masses formed massive gas clouds that eventually collapsed under their own gravity, igniting nuclear fusion and giving birth to stars.

When massive stars exhaust their fuel, they explode in a supernova, scattering newly formed elements—everything from carbon to uranium—across space. These elements coalesce again under the influence of gravity, forming planets, moons, and rocky worlds like our own. Smaller stars, like our Sun, do not explode but eventually burn out and settle into a dense, inert state—an eventual fate for our solar system (see Appendix C).

As this illustrates, gravity is the architect of the cosmos—the force responsible for the existence of stars, planets, and life itself. But what if gravity had been just a tiny bit different?

- If gravity were slightly weaker, atoms would never have clumped together. No stars, no planets, and no chemistry would have formed.
- If gravity were slightly stronger, atoms would have collapsed into a single dense mass shortly after the Big Bang. Again, no stars or planets—just one massive, lifeless object.

But how slight is "slight"?

To understand this, let us grasp the scale of measurement:

- A centimeter is one hundredth of a meter.
- A millimeter is one tenth of a centimeter.
- A nanometer is one millionth of a millimeter.
- A yoctometer is one septillionth of a meter: 1 meter ÷ 1,000,000,000,000,000,000,000,000.

Now, here is the astonishing part: The gravitational constant—the value used in the formula that calculates the force of gravity[44]—must be so precise that even a change as small as one part in a yoctometer would make the universe uninhabitable.

- If that constant were just slightly smaller, gravity would be too weak to form stars and galaxies.
- If it were just slightly larger, matter would collapse too quickly into singularities before anything could form.

---

[44]The gravitational force between two masses is described by the formula $F = (G \times m_1 \times m_2) / d^2$, where m_1 and m_2 represent the masses of the two objects in kilograms, d^2 is the square of the distance between them in meters, and G is the universal gravitational constant.

Only one incredibly narrow range of values allows for a universe that can support complex structures—and life.

This same principle applies to the strong nuclear force. Though its range is even smaller—less than a billionth of a millimeter—its strength must also be finely tuned. If it were changed by just a billionth of a yoctometer, atomic nuclei could not form and matter itself would not exist.

That is how delicate the balance is. When we use the word "slightly" in the context of the physical constants of the universe, we are referring to changes so minuscule that they stretch the limits of comprehension. Yet those infinitesimal differences determine whether the universe exists—or collapses into nothingness.

So, we must ask:

Is this extraordinary precision the result of random chance? Or does it point to something more—a purposeful design? Coincidence? Luck? Or something greater?

As of the time of writing, scientists have identified at least ninety-three known forces, constants, proportions, velocities, and distances that govern the formation and preservation of all matters in the universe. Each of these values is set with extraordinary precision. It is precisely because of their current, exact values that the universe is stable, structured, and capable of supporting life. Even the slightest deviation in any one of these constants would disrupt the behavior of matter and render the universe uninhabitable.

This raises a fundamental question: Could chance have produced the exact values necessary for everything we observe to exist? The probabilities involved are not merely improbable—they are astronomically implausible.

Take, for instance, the gravitational constant, which determines the strength of gravity. According to physicists, only one value in 10^{279} possible options would result in a universe capable of forming stable atoms—the foundational building blocks of matter and life (see Appendix B). In other words, the probability that gravity alone would have the right value by chance is 1 in 10^{279}.

Or consider the cosmological constant, which governs the rate of expansion of the universe. For the universe to expand at just the right rate—not too quickly to prevent matter from clumping together, and not too slowly to cause it to collapse—only one in 10^{57} possible values will suffice.

To illustrate this staggering improbability, astrophysicist Trinh Xuan Thuan[45], in his book *Le Chaos et l'Harmonie*, offers a memorable analogy:

> That number is so small that it corresponds to the probability that an archer would hit a 1 cm^2 target located at the other end of the universe, blindly shooting a single arrow from Earth and not knowing in which direction the target is.

Even more, when it comes to the strong nuclear force, physicists John Barrow and Frank Tipler[46] estimate that the probability of it having the precise value it does is 1 in 10^{32}. Now, if we calculate the joint probability that just these three fundamental forces—gravity, the cosmological constant, and the strong nuclear force—simultaneously possess the exact values needed for life to exist, the combined probability is 1 in 10^{368}.

And keep in mind—that is just three out of the ninety-three known constants. We have not even accounted for the remaining ninety variables.

To put this into perspective: one of the most well-known lotteries in the world, the Powerball in the United States, requires matching five numbers out of sixty-nine plus one "Powerball" out of twenty-six. The odds of winning? 1 in 292,201,338, or roughly 1 in 2.92 x 10^8—a probability we consider extremely remote.

---

[45]Trinh Xuan Thuan (born August 20, 1948, in Hanoi) is a Vietnamese-American astrophysicist and author who writes in French. He was awarded the UNESCO Kalinga Prize in 2009 and the Cino Del Duca World Prize in 2012. Among his notable works is the book *Le Chaos et l'Harmonie*, in which he explores the numerical basis for the possible values of the cosmological constant related to the universe's rate of expansion.

[46] *The Anthropic Cosmological Principle.*

But compared to the odds of 1 in 10^{368}, the Powerball jackpot begins to look almost guaranteed.

To attribute the precise calibration of the universe's physical constants to random chance is to suggest a coincidence so vast, so wildly improbable, that it borders on the mathematically absurd. Far from a rational explanation, it becomes a leap of blind faith—one that ignores the overwhelming evidence of fine-tuning at the heart of the cosmos.

Matter and the forces that govern it appear to have been designed from the very beginning with precisely defined properties. This extraordinary precision suggests that the universe followed a blueprint— a design laid out by a Designer. To claim that such exact calibration arose purely by chance is not merely speculative—it would be the greatest leap of blind faith one could make.

As scientists began uncovering the extraordinary fine-tuning of the universe—forces, constants, ratios, velocities, and distances—many in the believing community found their convictions reinforced by nothing less than science itself. The smallest variation in these values would render the universe impossible. Such facts point convincingly to the existence of a superior intelligence—a Creator—who determined the physical laws with such harmony and precision that the formation of matter, stars, planets, and life became possible.

These revelations leave no room for randomness. The equation that governs our cosmos carries not the fingerprints of chaos, but the signature of purpose.

Even the world-renowned physicist Stephen Hawking[47], not known for endorsing theism, acknowledged the mystery. In his 1988 classic, A Brief History of Time, he wrote:

> The laws of science, as we know them at present, contain many fundamental numbers, like the size of the electric charge of the

---

[47] Stephen William Hawking (1942–2018) was a British theoretical physicist, cosmologist, and science communicator, renowned for his work on the origins and structure of the universe, particularly in the fields of black holes and cosmology. He was also known for engaging in discussions on the relationship between science and religion, including arguments against the necessity of a divine creator based on scientific reasoning.

electron and the ratio of the masses of the proton and the electron... The remarkable fact is that the values of these numbers seem to have been finely adjusted to make possible the development of life.

Similarly, Fred Hoyle[48], the esteemed British mathematician, physicist, and astronomer—himself an agnostic—confessed:

> A commonsense interpretation of the facts suggests that a super intellect has monkeyed with physics, as well as with chemistry and biology, and that there are no blind forces worth speaking about in nature. The numbers one calculates from the facts seem to me so overwhelming as to put this conclusion beyond question.

Faced with such compelling implications, many atheist academics responded swiftly. Enter the theory of the multiverse[49]. Borrowed from the fringes of science fiction, this idea proposes that our universe is just one of trillions generated every second in a hypothetical "universe factory." Each of these universes supposedly has different laws, constants, and parameters. Most are failures—disintegrating instantly due to unstable conditions—but by sheer statistical chance, ours happens to have the right values for life.

And what evidence supports the existence of this cosmic factory? None. Not even within the most speculative boundaries of science fiction did the concept hold such elevated status as it does now in some academic circles.

But here is the deeper issue: even if such a "factory" existed, it does not solve the problem—it only pushes it back a step. Where did the factory come from? What forces and constants allowed it to operate? What raw materials did it use? What intelligence programmed it to test different values and produce functioning universes?

---

[48]Fred Hoyle was responsible for one of the most significant discoveries of the 20th century: carbon nucleosynthesis. He was an active member of both the Royal Society and the American Academy of Arts and Sciences. Hoyle passed away in 2001.

[49]You can watch world-renowned naturalist Richard Dawkins explain this theory in the following video: https://www.youtube.com/watch?v=oO0QRUX4HGE

In the past, the origin of the universe was traced back to a single, mysterious "ball" of energy from which the Big Bang erupted. For believers, this origin was attributed to a Creator, who encoded the necessary laws and properties into matter. Nonbelievers, on the other hand, claimed this initial state had simply always existed, and that chance was responsible for everything that followed.

However, the overwhelming fine-tuning we observe today forced a shift: chance could no longer explain the current state of the universe. Thus, the multiverse theory was born—not from observation, but from the need to preserve a worldview without design.

Ironically, this new theory leads to the same kind of puzzle. The question once was: "Where did matter come from?" Now, it is: "Where did the factory come from?" And if even one universe strains our understanding, how much more incomprehensible would a mechanism capable of producing infinite universes be?

Once again, there is no answer.

FIFTH THESIS: A PLANET OUT OF THE ORDINARY

It is a common theme in science fiction for extraterrestrial beings to visit Earth—often with hostile intent, seeking to destroy us for no clear reason, driven by desperation to acquire a vital natural resource that is scarce on their home planet but plentiful here. Some stories portray these visitors as highly evolved intelligence, observing us as scientists study lab mice, hoping to better understand what they themselves were like in a distant evolutionary past. Others, however, imagine aliens arriving simply to befriend us—integrating into human society, forming relationships, even marrying Earthlings, and establishing a foothold for their kind on our planet.

With the rise of such narratives, the idea that life might be as widespread as the stars and planets themselves began to take hold in the public imagination. Mars became one of the earliest subjects of speculation. People envisioned it as home to beings far more advanced than us—creatures capable of building spacecraft that could traverse millions of kilometers to reach Earth. The term "Martians" was coined to describe these imagined inhabitants.

English author H.G. Wells popularized this idea in his groundbreaking 1898 novel *The War of the Worlds,* in which he depicted a failed Martian invasion of Earth. The Martians, despite their technological superiority, succumbed to Earth's microscopic organisms, having no immune defenses against our common bacteria. The success of Wells's work sparked a cultural fascination with extraterrestrial life that persists to this day, continually invigorated by new discoveries in astrophysics and space exploration.

Voyager 1, a spacecraft launched from Cape Canaveral, Florida, on September 5, 1977, was originally designed for a mission lasting about twenty years. Yet, defying expectations, it continues its voyage to this day, now drifting through the vastness of interstellar space on a path toward the center of our galaxy.

Among the many remarkable transmissions from Voyager 1, perhaps none is more profound than the iconic "Pale Blue Dot"[50] photograph. This image—an unassuming yet deeply evocative portrait of our planet— stands as one of the most important visuals ever captured by humanity. However, it is not the kind of picture we associate with views from the Moon or the International Space Station, where Earth appears as a vibrant sphere with discernible continents, swirling clouds, and vast blue oceans.

Instead, this photograph reveals Earth as nothing more than a tiny speck—barely distinguishable, comparable in size to the tip of a pin— suspended in a ray of scattered sunlight. Taken from a staggering distance of six billion kilometers away (for comparison, the average distance between Earth and the Sun is 150 million kilometers), the image was captured on February 14, 1990. It serves as a humbling reminder of our planet's fragility and insignificance against the vast backdrop of the cosmos. In response to this photograph, astronomer Carl Sagan[51]

---

[50]https://voyager.jpl.nasa.gov/galleries/images-voyager-took/solar-system-portrait/

[51]Carl Edward Sagan (New York, November 9, 1934 – Seattle, December 20, 1996) was an American astronomer, astrophysicist, cosmologist, astrobiologist, author, and science communicator. He was a strong advocate of scientific skepticism and the scientific

published *A Pale Blue Dot: A Vision of the Human Future in Space,* four years later. One of the book's chapters states:

> Look again at that dot. That's here. That's home. That's us. On it everyone you love, everyone you know, everyone you ever heard of, every human being whoever was, lived out their lives. The aggregate of our joy and suffering, thousands of confident religions, ideologies, and economic doctrines, every hunter and forager, every hero and coward, every creator and destroyer of civilization, every king and peasant, every young couple in love, every mother and father, hopeful child, inventor and explorer, every teacher of morals, every corrupt politician, every "superstar," every "supreme leader," every saint and sinner in the history of our species lived there-on a mote of dust suspended in a sunbeam.
>
> The Earth is a very small stage in a vast cosmic arena. Think of the endless cruelties visited by the inhabitants of one corner of this pixel on the scarcely distinguishable inhabitants of some other corner, how frequent their misunderstandings, how eager they are to kill one another, how fervent their hatreds. Think of the rivers of blood spilled by all those generals and emperors so that, in glory and triumph, they could become the momentary masters of a fraction of a dot.
>
> Our posturing, our imagined self-importance, the delusion that we have some privileged position in the Universe, are challenged by this point of pale light. Our planet is a lonely speck in the great enveloping cosmic dark. In our obscurity, in all this vastness, there is no hint that help will come from elsewhere to save us from ourselves.
>
> The Earth is the only world known so far to harbor life. There is nowhere else, at least in the near future, to which our species could migrate. Visit, yes. Settle, not yet. Like it or not, for the moment the Earth is where we make our stand.
>
> It has been said that astronomy is a humbling and character-building experience. There is perhaps no better demonstration of the folly of human conceits than this distant image of our tiny world. To me, it underscores our responsibility to deal kindlier with one another, and to preserve and cherish the pale blue dot, the only home we've ever known.

---

method, a pioneer in the field of exobiology, and a key promoter of the search for extraterrestrial intelligence through the SETI (Search for Extraterrestrial Intelligence) project.

The SETI (Search for Extraterrestrial Intelligence) project is dedicated to the quest for intelligent life beyond Earth. It does so by analyzing electromagnetic signals—such as radio waves, television broadcasts, mobile phone transmissions, and even light emitted by streetlamps—captured by various radio telescopes. In addition to passively listening, SETI also engages in active efforts by sending different kinds of messages into space, hoping that one day, one might receive a response.

If intelligent beings exist elsewhere in our galaxy and are conducting a search like ours, they will need to be located within approximately one hundred light-years of Earth to detect any of our earliest signals. This is the estimated distance that some of the first terrestrial broadcasts, like those from the BBC in London beginning in 1922, have traveled. While 100 light-years—equivalent to about 9.4×10^{14} kilometers—is an immense distance by human standards (see Appendix B), it is relatively small when compared to the vast span of our Milky Way galaxy, which measures roughly 100,000 light-years across (or 9.4×10^{17} kilometers). In galactic terms, any civilization capable of hearing our signals would have to be in our cosmic neighborhood—like living on the same city block.

SETI is not the only initiative of its kind. Numerous other projects, both in the United States and across Europe, are engaged in the search for extraterrestrial intelligence. To date, however, no definitive signals of intelligent origin have been detected. Still, as the saying goes, "absence of evidence is not evidence of absence." We cannot yet conclude that intelligent life beyond Earth does not exist—and perhaps we never truly will be able to.

In 1950, Italian physicist Enrico Fermi—Nobel laureate and widely recognized as the "father of the nuclear reactor"—formulated what is now known as the Fermi Paradox. This paradox highlights the apparent contradiction between the high probability of intelligent life existing elsewhere in the universe and the complete absence of any concrete evidence confirming its existence.

Over the past seventy years, our understanding of the cosmos has expanded dramatically. Much of this progress is due to technological advancements that have allowed us to overcome the visual limitations imposed by Earth's atmosphere, which acts like a hazy, semi-transparent veil that distorts our view of the universe. The deployment of space-based observatories—most notably the Hubble Space Telescope[52], launched on April 24, 1990—has enabled us to peer deep into space with unprecedented clarity, free from atmospheric interference.

Given how little was known in 1950, it is not surprising that scientists of the time speculated that numerous planets like Earth might exist—worlds capable of supporting complex life.

To estimate the likelihood of life existing elsewhere in the universe—even if it is not intelligent—we must first identify the minimum requirements a planet must meet to support life. Once these essential conditions are defined, we can then assess the probability of such planets existing elsewhere in space.

For centuries, humanity held a geocentric view of the cosmos, believing that Earth was the center of the universe and that the Sun, planets, and all celestial bodies revolved around it. This worldview was profoundly challenged by Nicolaus Copernicus[53], who spent 25 years developing his seminal work *De Revolutionibus Orbium Coelestium* (On the Revolutions of the Celestial Spheres), completed in 1532. In it, Copernicus demonstrated that the Sun, not Earth, lies at the center of our solar system, and that Earth, along with the other planets, revolves around it. This heliocentric model marked a turning point in scientific thought and forever altered our understanding of our place in the cosmos.

With this revolutionary insight, Earth lost its perceived central and privileged position. Instead of being the focal point of the universe, our planet was reclassified as just one among many orbiting a common star.

---

[52]https://www.nasa.gov/mission_pages/hubble/main/index.html

[53]Nicolaus Copernicus (February 19, 1473 – May 24, 1543) was a Polish Renaissance astronomer and canon who formulated the heliocentric model of the solar system—though the concept had been previously proposed by Aristarchus of Samos. His seminal work, *De revolutionibus orbium coelestium* (On the Revolutions of the Celestial Spheres), is widely regarded as the foundational text of modern astronomy and a cornerstone of the Scientific Revolution during the Renaissance. The book was published posthumously in 1543 by Andreas Osiander.

Over time, this perspective gave rise to what became known as the Copernican Principle—the idea that Earth and its circumstances are not unique or special. If life arose here, under certain conditions, it stands to reason that similar conditions could exist—and similar life could arise—elsewhere.

Further expansion of our cosmic understanding came in 1921, when astronomer Edwin Hubble[54] discovered that many of the bright spots in the night sky previously thought to be stars were in fact entire galaxies, each composed of billions of stars, planets, and other celestial bodies. Until then, it was widely believed that the Milky Way encompassed the entire universe. Hubble's revelation shattered that notion, expanding the known universe by unimaginable scales—trillions upon trillions of times larger than once believed.

With this newfound perspective, the idea that planets like Earth might be common gained scientific credibility. If galaxies are filled with billions of stars, and many of those stars host planets, then the potential for Earth-like worlds—capable of harboring life—could be vast. Under the Copernican Principle, life is not a cosmic miracle confined to a single world, but rather a phenomenon that might be woven throughout the fabric of the universe.

> I imagine that the number of inhabited planets in our galaxy is of the order of thousands or hundreds of thousands. And why do I think there is life on other planets? Because the universe is extremely large, there are billions and billions of stars. So, unless our Earth has something very special, very special, miraculous if you will, what has happened here on Earth must have happened many times on other planets. (Seth Shostak, Senior SETI Astronomer)

The hypothesis that life may exist beyond Earth gave rise to the scientific field of astrobiology—a discipline devoted to understanding the

---

[54]Edwin Powell Hubble (Marshfield, Missouri, November 20, 1889 – San Marino, California, September 28, 1953) was one of the most prominent American astronomers of the 20th century. He is best known for what was long believed to be his 1929 demonstration of the universe's expansion—a discovery that fundamentally changed our understanding of the cosmos. Hubble is regarded as the father of observational cosmology, although his influence extends across many areas of astronomy and astrophysics.

origins, evolution, distribution, and potential future of life in the universe. One of its primary missions is to determine whether habitable planets are rare exceptions or common occurrences in the cosmos.

A leading contributor to this field is Guillermo González[55], an astrobiologist and astrophysicist at Iowa State University, who works closely with NASA's astrobiology programs. The overarching goal of such research is to identify the key conditions necessary for life and to determine whether these conditions exist elsewhere in the universe.

Astrobiology rests on two foundational assumptions. First, there are millions upon millions of stars in the universe, many of which are accompanied by planetary systems. Second, that the emergence and persistence of complex life requires an intricate chain of events and extremely precise environmental conditions. Among the most critical of these is the presence of liquid water, which is essential for all known forms of life.

For liquid water to exist, a planet must orbit its star at just the right distance—not too close and not too far. If it is too close, water would evaporate due to extreme heat; too far, and water would freeze solidly. This narrow orbital range is referred to as the "Goldilocks zone[56]"—a region within each solar system where conditions are just right for liquid water to exist on a planet's surface.

In our own solar system, Earth occupies this precise zone. Scientists estimate that if Earth were just 5% closer to the Sun, it would experience a runaway greenhouse effect like that of Venus, with surface temperatures soaring to around 900°F, rendering the planet uninhabitable and devoid of liquid water. Conversely, if Earth were 20%

---

[55]Guillermo González (born 1963 in Havana, Cuba) is an astrophysicist known for advocating the principle of intelligent design. He serves as an assistant professor at Ball State University in Muncie, Indiana. González is also a senior fellow at the Discovery Institute's Center for Science and Culture—considered a central hub of the intelligent design movement—and a member of the International Society for Complexity, Information and Design, which likewise promotes intelligent design.

[56]"This oatmeal is too hot," Goldilocks exclaimed. So, she tried the oatmeal from the second bowl. "This oatmeal is too cold," she said. Then she tried the last bowl of oatmeal. "Ah, this oatmeal is just right!" she said happily and ate it all. (Excerpt from the children's story Goldilocks and the Three Bears)

farther away, it would resemble Mars—a cold desert world where carbon dioxide clouds dominate the atmosphere and water would freeze, making life as we know it impossible.

Because the laws of physics and chemistry are universal, scientists base their search for habitable exoplanets on these same principles. Accordingly, efforts to identify alien worlds capable of supporting life focus primarily on finding planets situated within their star's Goldilocks zone—places where liquid water, and potentially life, might thrive.

While the presence of liquid water is essential for life, it is far from the only requirement. The recipe for a life-supporting planet is profoundly complex, involving a delicate interplay of astrophysical, geological, and atmospheric conditions. For a planet to harbor life— particularly complex life—it must meet a broad range of finely tuned criteria, including but not limited to the following:

- Just as solar systems have a "Goldilocks zone," galaxies have habitable zones as well. The core of a galaxy is typically a chaotic and dangerous region, marked by high stellar density, frequent supernovae, and intense radiation. Building a habitable planet there would be like constructing a home in a minefield surrounded by erupting volcanoes, within a tornado corridor, and atop shifting tectonic plates.
- The planet must orbit a G2-type[57] main-sequence star, like our Sun—a type that comprises only about 7.5% of stars in our galaxy. Smaller stars (e.g., red dwarfs) would require planets to orbit closer for warmth, which could lead to tidal locking, where one side of the planet always faces the star. This results in one hemisphere being a scorched desert and the other a frozen wasteland—conditions hostile to life as we know it.
- Massive gas giants, like Jupiter and Saturn, act as gravitational shields, attracting or deflecting comets, asteroids, and other potentially catastrophic objects that might otherwise impact Earth-like planets.

---

[57]For the classification of stars see https://es.wikipedia.org/wiki/Clasificaci%C3%B3n_estelar#Clase_G

- The planet must lie within its solar system's Goldilocks zone, the narrow band where temperatures allow liquid water to exist on the surface.

- A planet must have an orbit that is circular. Highly elliptical orbits would cause extreme temperature fluctuations—ranging from sub-freezing to nearly 1,000°F—that would make stable life impossible.

- An optimal atmosphere composed mainly of nitrogen (78%), oxygen (21%), and trace gases (such as carbon dioxide (1%)) is necessary to: Maintain a stable climate, Shield the surface from harmful solar wind, Enable the formation of liquid water, and Support aerobic respiration in complex organisms.

- The atmosphere must be sufficiently transparent to allow sunlight to penetrate to the surface, enabling photosynthesis—a process vital for generating oxygen and sustaining plant life.

- A large natural satellite, such as our Moon (about 25% the size of Earth), stabilizes the planet's axial tilt. This 23.5-degree tilt allows for moderate seasonal variation and a consistent 24-hour day-night cycle. Without the Moon[58], the tilt could vary wildly—between 0° and 90°—resulting in catastrophic climatic instability and rapid rotational speeds.

- The planet must have a magnetic field generated by a liquid iron core, which protects the surface from solar radiation and charged particles that could otherwise strip away the atmosphere.

- A planet must have adequate mass to retain its atmosphere and maintain a strong magnetic field. If it were significantly smaller, like Mars, the planet would be vulnerable to atmospheric loss and surface desiccation.

- An ideal ratio of approximately 70% water to 30% land fosters biodiversity, regulates global temperatures, and supports a variety of ecosystems and weather patterns essential for sustaining life.

- The planet must rotate at a moderate speed. Too fast, and it becomes a heat-trapping furnace; too slow, and the temperature

---

[58]See https://elpais.com/elpais/2015/12/15/ciencia/1450179769_533306.html

contrast between day and night becomes too extreme for life to adapt.

- The thickness of the planet's crust (Earth's ranges from 4 to 30 miles) is critical. If it is too thick, plate tectonics cannot occur; if too thin, no stable landmass will form. Plate tectonics[59] help regulate the global climate, recycle nutrients, and facilitate the carbon cycle, which is essential for organic chemistry and the development of life's molecular building blocks.

Taken together, these conditions illustrate the extraordinary degree of fine-tuning required for a planet to support life. While the universe is vast and diverse, the convergence of all these parameters on a single world may be exceptionally rare, making planets like Earth truly precious.

For complex life to develop and endure, a remarkably specific and interconnected set of conditions must be met—not in isolation, but concurrently. Over time, the number of criteria considered essential for a planet to be habitable has grown. Current estimates suggest that at least twenty[60] critical requirements must be satisfied for a planet to support life as we know it.

To understand the rarity of such a planetary configuration, let us consider a conservative estimate: suppose that there is a 1 in 10 chance

---

[59]This recycling occurs through the movement of the Earth's outermost layer, which is broken into massive sections, much like pieces of a puzzle. These tectonic plates slide over and under one another in a process known as subduction. As one plate is forced beneath another, it sinks toward the Earth's core and begins to melt. Meanwhile, new crust forms elsewhere, continuing the cycle of plate formation and renewal.

[60]The Drake Equation, also known as Drake's formula, is used to estimate the number of civilizations in our galaxy (the Milky Way) that might be capable of emitting detectable radio signals. It was formulated in 1961 by radio astronomer Frank Drake—then working at the National Radio Astronomy Observatory in Green Bank, West Virginia—while serving as president of the SETI Institute. The equation outlines several key factors believed to influence the emergence and detectability of extraterrestrial civilizations. Although current data are insufficient to produce a definitive solution, the scientific community recognizes the equation as a valuable theoretical framework. It has served as a foundation for numerous hypotheses regarding the existence of intelligent life beyond Earth. You can view the full formula and its explanation here: https://en.wikipedia.org/wiki/Drake_equation

(10%) for any given planet to meet each of these individual requirements. If these criteria are statistically independent, the probability of a single planet meeting all twenty is: $(1/10)^{20} = 1$ in 1×10^{15}

Now, consider the estimated number of stars in our galaxy: approximately 1×10^{11}. If each star hosts just one planet within its habitable or "Goldilocks" zone (a generous assumption), that gives us roughly 1×10^{11} potential candidates.

But here is the staggering conclusion: if the probability of meeting all twenty conditions is 1 in 10^{15}, and there are only 10^{11} available planets, then the likelihood of even one planet meeting all the criteria is effectively less than one. In other words, we should not be here—statistically speaking.

To illustrate this, imagine playing the Powerball lottery with sixty-nine main numbers and 26 Powerball options. Now suppose that 90% of all possible number combinations are never selected by any player. For someone to still win under those conditions it would be considered nothing short of miraculous. And yet, that is precisely the sort of improbability our existence represents.

This leads to a provocative philosophical question:

Are we merely lucky? Or was everything predetermined?

Adding to this mystery is the remarkable precision required for total and flawless solar eclipses to occur—phenomena that have played a pivotal role in advancing our understanding of the universe. In 1919, a total solar eclipse famously provided the first observational confirmation of Einstein's General Theory of Relativity, marking a revolutionary leap in modern physics. Eclipses have also given scientists critical insights into the structure of the sun, including its corona, which emits solar wind, ultraviolet radiation, and heat.

But such eclipses are extraordinarily rare on a cosmic scale. A total eclipse of the Sun occurs only when the apparent size of the Moon perfectly matches the apparent size of the Sun, as seen from Earth. This is possible only because: The Sun is 400 million times larger than the Moon and the Moon is 400 million times closer to Earth than the Sun.

A deviation greater than 2% in this delicate ratio would render total eclipses either too complete (blocking the corona entirely and providing no useful data) or too incomplete (allowing too much sunlight to pass through, overwhelming instruments and visibility). In Astronomy and Geophysics, Guillermo González notes that of all the moon-bearing planets studied, only Earth meets the precise conditions required to observe perfect total solar eclipses.

Again, the question arises: Are we simply fortunate, or is our existence part of a grand design?

In their groundbreaking work *Rare Earth: Why Complex Life is Uncommon in the Universe*, scientists Peter Ward and Donald Brownlee delve deeply into these questions. Their research concludes that while microbial life may be common throughout the cosmos, the emergence and sustainability of complex life—such as plants, animals, or intelligent beings—is exceedingly rare. The environmental, astronomical, and biochemical requirements are so strict and specific that Earth may indeed be a singular oasis in an otherwise inhospitable universe.

So, we return, once more, to the hauntingly beautiful dilemma: Are we lucky? Or was everything predetermined?

SIXTH THESIS: THE LAWS OF THERMODYNAMICS

Thermodynamics is the scientific study of energy and its transformations. Its origins trace back to the mid-nineteenth century, a time when what we now call "energy" was commonly referred to as "force." The first two principles of thermodynamics, originally formulated with respect to "closed systems," form the foundation of modern scientific thought.

A closed system is one that does not exchange matter with its environment, though it may still transfer energy in the form of heat or work. For example, if we analyze a liquid contained in a hermetically sealed vessel that is completely insulated from external influences like air, light, and temperature, then that vessel is considered a closed system. If we expand this concept by sealing off an entire laboratory—ensuring no external light, air, or sound can enter—the laboratory itself becomes a closed system for the purpose of studying the materials within.

The first law of thermodynamics, also known as the law of conservation of energy, states that within a closed system, energy is neither created nor destroyed—it merely changes form. Energy remains constant, though it may transform repeatedly. Consider the example of burning a piece of wood. The wood is transformed into ash, releasing energy in the form of heat and light. This energy did not come from outside the system; it was part of it all along. After combustion, the total energy remains the same, albeit in different forms. While proving this rigorously lies beyond the scope of this discussion, it is a well-established principle in any standard physics text.

The second law of thermodynamics, commonly referred to as the law of entropy[61]—and once described by Albert Einstein[62] as the "supreme law of all science"—states that, over time, the natural tendency in a closed system is for disorder, or entropy, to increase. This tendency can be observed everywhere. Objects wear out, decay, rust, break, disintegrate, rot, or fade. Complex and orderly matter gradually becomes simple and disordered. Think of the human body, which is intricate and highly organized. What becomes of it after two centuries? It eventually breaks down into a disorganized, dusty residue—ashes. Can this process of decay be stopped or reversed? Not within a closed system. However, if we expand the system—allowing external energy to be introduced—we can interfere with this process. For instance, if you were to lock up your house and leave it untouched for many years, you would return to find it in disrepair: dust, broken fixtures, possible structural collapse. Left to itself, the house cannot repair the damage. But if you open the system by re-entering it and applying energy in the form of labor and resources, you can restore it. The input of external energy changes the equation.

---

[61]Entropy is a measure of the disorder or randomness in a system, particularly in the arrangement and motion of gas molecules. It is often associated with concepts such as chaos, unpredictability, and molecular disorganization.

[62]Albert Einstein (1879–1955) was a German physicist of Jewish origin, widely regarded as the most influential, well-known, and iconic scientist of the 20th century. In 1921, he was awarded the Nobel Prize in Physics for his explanation of the photoelectric effect, which was pivotal to the development of quantum theory.

This law also complements the first. While the total amount of energy in a closed system remains unchanged, as the first law dictates, the second law explains that this energy becomes progressively less "useful." Revisit the example of the burned wood. The ash, heat, and light produced are all forms of energy, but they are less reusable than the original unburned wood. You could attempt to recombine the remnants, but the energy output from a second combustion would be far lower than the first. With each successive transformation, the amount of usable energy decreases until it is no longer possible to extract meaningful work from it.

In the real world, entropy cannot be avoided, but it can be mitigated. This is the essence of engineering: designing systems that make better use of energy before it becomes too degraded to be of practical value. That is why terms like "efficiency" are so central in technical disciplines. When one motor is said to be more efficient than another, it means more of the input energy is being converted into useful work—a smaller portion of it is lost as heat, vibration, or noise. Still, according to the second law of thermodynamics, no machine can be perfectly efficient. Some portion of energy will always be lost in non-recoverable forms.

Energy persists, but its usefulness fades. This is the profound and unavoidable truth that thermodynamics teaches us—and it has implications that extend from engineering to everyday life.

The first and second laws of thermodynamics are often brought into discussions about the origin of the universe, particularly in debates over the existence of God. If we reject the notion of a Creator, the existence of the universe must still be rationally explained. According to the prevailing scientific narrative—commonly summarized in the "big story" (see Appendix C)—the universe began with a tremendous release of energy from an extremely dense and minuscule "ball" that suddenly exploded, initiating the expansion of space, time, and matter. This raises a fundamental question: What is the origin of that original "ball" of energy?

There are essentially three possible explanations: spontaneous generation (something from nothing), eternal existence (matter has always existed), or intentional creation (matter was brought into existence by an external agent).

The first hypothesis—spontaneous generation—stands in direct conflict with the first law of thermodynamics, which states that energy in a closed system cannot be created or destroyed, only transformed. If absolutely nothing existed prior to this supposed explosion, there would have been no energy or matter available to transform. According to the first law, the spontaneous appearance of matter and energy from literal nothingness is not scientifically permissible. Thus, invoking spontaneous generation would require a force or influence outside the closed system of the universe—something that could introduce energy and matter into it. For Christians, this aligns with the belief in a transcendent Creator: The God of the Bible, who exists outside of time, space, and matter, and who initiated the existence of all things.

The second hypothesis—that the universe has existed eternally—runs into direct conflict with the second law of thermodynamics, which affirms that in any closed system, the amount of usable energy inevitably decreases over time. If the universe had no beginning and has existed for an infinite duration, it should have long ago reached a state of maximum entropy—a state in which all usable energy is depleted, and no further work or change would be possible. Yet we observe that the universe is still in a state of dynamic activity: stars are forming, galaxies are evolving, and life continues to emerge and develop. This observable reality implies that not all usable energy has been expended, suggesting that the universe has not existed forever. Therefore, the second law undermines the plausibility of an eternal universe.

After dismissing the first two hypotheses—spontaneous generation and eternal existence—as incompatible with the foundational laws of thermodynamics, only one rational explanation remains: creation by an intelligent Creator. This conclusion is not merely a product of theological conviction but emerges from a logical analysis of physical reality itself. The existence of the universe and life, within the constraints of scientifically established principles, points beyond itself to a cause that is not bound by those principles—a cause outside the closed system of nature.

Even some prominent scientists have acknowledged this implication. Robert Jastrow[63], a renowned astronomer, physicist, and founder of NASA's Goddard Institute for Space Studies, famously wrote:

> The essence of the strange developments is that the Universe had, in some sense, a beginning—that it began at a certain moment in time, and under circumstances that seem to make it impossible—...*Theologians generally are delighted with the proof that the Universe had a beginning, but astronomers are curiously upset...Now we see how the astronomical evidence leads to a biblical view of the origin of the world. The details differ, but the essential elements and the astronomical and biblical accounts of Genesis are the same; the chain of events leading to man commenced suddenly and sharply at a definite moment in time, in a flash of light and energy.* ... Consider the enormity of the problem. Science has proved that the universe exploded into being at a certain moment. It asks: What cause produced this effect? Who or what put matter or energy into the universe? ... There is a kind of religion in science. ... This religious faith of the scientist is violated by the discovery that the world had a beginning under conditions in which the known laws of physics are not valid, and as a product of forces or circumstances we cannot discover. When that happens, the scientist has lost control. If he really examined the implications, he would be traumatized.[64] (*emphasis* mine)

"The Big Story" (Appendix C) has captured widespread interest, not only because it attempts to trace humanity's origins back to the very beginning of time, but also because it dares to address some of the most profound and complex questions about existence. Among these are: If, as the second law of thermodynamics states, the universe tends toward disorder and disintegration, how could such a vast and highly ordered cosmos have emerged at all? How has life—so structured, dynamic, and purposeful—appeared in defiance of this natural tendency toward

---

[63]Robert Jastrow (1925–2008) was an American scientist who made significant contributions in the fields of astronomy, geology, and cosmology. He authored numerous popular science books and articles, helping to bring complex scientific ideas to a general audience. In 1961, he founded NASA's Goddard Institute for Space Studies. He also served as director emeritus of the Mount Wilson Observatory and was a professor at Columbia University, where he earned his Ph.D. in Theoretical Physics. Jastrow was widely regarded as one of the leading astrophysicists of his time.

[64] *God and the Astronomers.*

entropy? How could it have steadily advanced through a chaotic universe to reach its current splendor? And most provocatively, how could a colossal explosion, like the hypothesized Big Bang, initiate a chain of increasingly ordered events that culminate in galaxies, planets, ecosystems, and conscious beings?

It is worth recalling that the original "ball" of energy in the Big Bang theory is assumed to have contained all the raw materials necessary to form everything in the universe. Yet, according to the second law of thermodynamics, explosions are inherently chaotic. They lead to disarray, not structure. This makes the standard narrative difficult to reconcile with the observed reality of increasing complexity. To draw an analogy, imagine placing gears, screws, shards of glass, metal fragments, and a lit stick of dynamite inside a sealed jar—and the resulting explosion producing a finely tuned, ticking wristwatch. Such a notion is not just improbable; it is counter to everything we know about how disorder behaves.

Naturalists—those who maintain that nature is the sole reality—attempt to explain all phenomena strictly within the boundaries of natural laws and physical causes. For them, only what can be observed, measured, and tested in a laboratory is real. Supernatural explanations are, by definition, excluded. By contrast, theistic belief embraces the existence of realities beyond the physical, holding that both the natural and the supernatural originate from an intelligent Creator who transcends the material world.

How, then, might an educated naturalist respond when faced with the tension between the second law of thermodynamics and the apparent rise of order, life, and complexity in the universe? In most cases, they would acknowledge that science does not yet have a definitive answer. This was precisely the response given by Michael Shermer[65], founder of *The Skeptics Society* and editor-in-chief of *Skeptic* magazine, when interviewed on *Faith Under Fire*, a program produced by Lee Strobel. Shermer admitted that, although the second law raises difficult questions, science has not yet resolved them.

---

[65]Michael Shermer holds a Ph.D. in the History of Science from Claremont Graduate University and is the author of numerous books on science and critical thinking.

On the other hand, those without extensive academic training often repeat what they learned through popular education: that matter, over immense spans of time, spontaneously organized itself—without intention or design—into simple organisms, which then evolved gradually into more complex and structured forms. This explanation is commonly accepted, though it glosses over the enormous improbabilities involved and the tension with the second law's principle of increasing disorder.

Yet the level of precision, structure, and timing observed at each key stage of cosmic and biological development suggests not a random series of accidents, but the involvement of intelligent direction. It points to a Great Designer who not only brought matter into existence but also instilled it with governing laws and directed its development at critical moments—those "turning points" in history where matter behaved in a novel way to ascend to a new stage of complexity. Such guidance implies intentionality, foresight, and planning—traits that align more naturally with theism than with blind material processes.

Even voices within secular scientific circles have acknowledged the strange tension posed by entropy. Evolutionist and social theorist Jeremy Rifkin[66] once remarked, "The law of entropy (or the second law of thermodynamics) will be the most paradoxical topic to be discussed in the next period of our history."

---

[66]Jeremy Rifkin is an American philosopher, scientist, economist, and political theorist. He is the author of numerous books, including *The Empathic Civilization*.

CONCLUSION

To support the case for the existence of God, I have argued that both matter and life display unmistakable signs of design—and where there is design, there must be a designer. That designer, as affirmed in Scripture, is the Creator: The God of the Bible.

A powerful illustration of how design signifies intelligence appears in the film *Contact*[67], where actress Jodie Foster portrays Dr. Ellie Arroway, a scientist working at the Arecibo Observatory in Puerto Rico. In the storyline, she spends years listening to random cosmic noise—until one day, she detects a meaningful signal: a sequence of pulses and pauses that correspond precisely to the first twenty-five prime numbers[68] (from 2 to 101). The pattern begins with two pulses, then a pause; followed by three pulses, a pause; then five pulses, and so on. It is immediately clear to the character—and to any thinking observer—that this is not a random occurrence. The presence of a mathematical pattern, one that specifically mirrors the sequence of prime numbers, signals the involvement of an intelligent source.

Why does she—and why do we—make that conclusion? Because there are no natural law compelling radio waves to arrange themselves into prime number sequences. Such complexity, specificity, and order are empirical indicators of purpose and planning. Just as a fingerprint at a crime scene strongly suggests a human presence, a structured and meaningful signal suggests an intelligent mind behind it.

This same logic applies—yet on a vastly grander scale—to the world around us. From the microscopic intricacies of cellular machinery to the breathtaking order of galaxies, we find hallmarks of design that are

---

[67] *Contact* is a 1997 American science fiction drama film directed by Robert Zemeckis. It is an adaptation of the 1985 novel of the same name, written by Carl Sagan.

[68] Prime numbers are those that can only be divided evenly by themselves and by one.

deeply consistent with the biblical account of creation. And yet, naturalists persist in embracing explanations that defy both common sense and scientific reasoning. They accept, without empirical support:

- That *nothing* produced *something*.
- That a hypothetical "multiverse machine" randomly produces billions of universes per second, each with varying physical constants, until one—ours—happens to emerge with exactly the values needed to sustain life.
- That life arose spontaneously from non-living matter, by chance alone.

What evidence undergirds these naturalistic hypotheses? Often, the only answer provided is *we are here*. Our existence is taken as the sole validation of these immense claims. But is that truly a scientific answer?

There are only two possibilities: either life was intentionally created, or life created itself through undirected natural processes. Naturalists overwhelmingly favor the latter—not because it is more reasonable or more supported by evidence, but because the alternative would force them to confront the existence of a Creator. Such an acknowledgment carries philosophical and theological implications that many are unwilling to accept.

But consider this: if a scientist heard a sequence of twenty-five prime numbers from space, they would not hesitate to conclude that intelligence was involved. By the fifth or sixth number, coincidence would be ruled out. By the tenth, intelligence would be all but certain. By the time the full sequence is heard, it would be impossible to deny it.

So why is the same logic not applied to the human genome, where we do not find a string of twenty-five, but three billion precisely ordered information units? Why does this not signal intelligence? Why do we ignore the staggering improbability—1 in 10^{368}—that the physical constants of the universe (such as gravity, the strong nuclear force, or the rate of cosmic expansion) would take on exactly the values required to allow matter, galaxies, stars, and life to form? Why is the near impossibility that only 1 in 10^{15} planets can support complex life not seen as evidence of deliberate design?

When evidence of intelligence is encountered in a radio signal, scientists affirm it without hesitation. But when the very structure of life, matter, and the universe speaks in the language of precision, order, and purpose, naturalists dismiss it as chance.

This is not a matter of rejecting science—it is a matter of applying scientific reasoning consistently. If we acknowledge that information and order points to intelligence in one domain, we must do the same across all domains. And when we do, the evidence strongly points to the conclusion that behind the universe stands not randomness, but reason; not chaos, but design; not blind chance, but the mind of a Creator.

Biogenesis is a foundational biological principle stating that life can only arise from pre-existing life. Despite this, some scientists have periodically proposed that life could have originated purely from inorganic matters. In the Middle Ages, it was widely believed that life—such as larvae and flies—spontaneously emerged from garbage and waste. But in 1668, Italian physician Francesco Redi[69] demonstrated that while these organisms *appeared* in garbage, they did not originate from it[70]; they came from eggs laid by other organisms.

Two centuries later, a new wave of scientists speculated again that microorganisms and algae might arise spontaneously from non-living material. In the mid-19th century, Louis Pasteur[71] put this hypothesis to

---

[69]Francesco Redi (Arezzo, February 18, 1626 – Pisa, March 1, 1697) was an Italian physician, naturalist, physiologist, and writer. He is regarded as the founder of helminthology, the scientific study of parasitic worms.

[70]To test his hypothesis, Redi placed a piece of meat into three identical jars. He left the first jar open, sealed the second with a cork, and covered the third with a tightly secured piece of cloth. After several weeks, he observed that larvae had appeared only in the open jar. Although the contents of the second and third jars had decomposed and produced a foul odor, no larvae were present. From this, Redi concluded that the flesh of dead animals does not spontaneously generate worms unless insect eggs are deposited in it.

To address the possibility that air might have influenced the outcome, he conducted a second experiment using meat and fish in a jar covered with fine gauze. Over time, he observed that flies did not enter the jar but laid their eggs on the gauze itself. The results mirrored those of his first experiment. From his work comes the famous phrase "*omne vivum ex ovo, ex vivo*", which translates as "all life comes from an egg, and that egg from the living."

[71]Louis Pasteur (Dole, France, December 27, 1822 – Marnes-la-Coquette, France, September 28, 1895) was a French chemist and bacteriologist whose groundbreaking

rest. Through a series of ingenious experiments—sealing sterilized broths in flasks or using swan-necked containers—he demonstrated conclusively that microorganisms only arise from other microorganisms. No life emerged where there was no prior life. Pasteur's work became a cornerstone of modern biology, confirming once again the law of biogenesis.

Yet a century after Pasteur, some scientists speculated that life on Earth might have spontaneously originated when atmospheric gases and chemical compounds encountered the right conditions. Space missions like *Viking I* and *Viking II*, launched in 1975 to explore Mars—believed to have had similar origins to Earth—sought evidence of past or present life that could support this idea. However, after extensive analysis, neither mission found any definitive signs of life.

Laboratory experiments over the years have succeeded in producing amino acids or organic molecules under controlled conditions. Occasionally, unusual chains of amino acids are synthesized, prompting headlines like *"Scientists Solve the Mystery of Life's Origin"* (as reported by *Phys.org*, August 1, 2019). Yet these headlines are often misleading. A few amino acids, some water on Mars, or the detection of organic material in meteorites do not equate to the creation of life. Life is not simply the presence of organic molecules; it is a highly organized, self-sustaining, information-driven system.

Even under ideal laboratory conditions—with carefully controlled temperature, light, and pH—placing all essential biological components (amino acids, carbohydrates, lipids) into a sterile container will not result in the spontaneous emergence of life. Remove all inhibitors, and still, life does not appear. Why? Because life is more than chemistry; it is structure, purpose, and information.

Suppose one day scientists do manage to generate life in a laboratory using the most basic building blocks. Such a feat would not prove spontaneous generation—it would instead prove *design*. It would confirm that life requires deliberate planning, vast intelligence, and highly precise execution. It would not refute God—it would affirm that

---

discoveries had a profound impact on the natural sciences, particularly in the fields of chemistry and microbiology.

the origin of life is an intelligent act. And if such an act were accomplished, it would merely replicate what has already been done—mirroring, not originating, the creative act of God.

Every serious experiment to date has reinforced the law of biogenesis. For naturalists to defend the spontaneous origin of life from non-living matter, they must reject this well-established law. In contrast, creationists remain consistent: life originates from life, and the original life came from the living God—the Creator who breathed life into matter. As Scripture says, "Let the land bring forth plants, those that produce seeds and fruit trees.", "Let the waters be filled with living creatures and let birds fly above the earth in the firmament of the heavens.", "Let the earth bring forth living creatures each according to its kind: cattle and reptiles and wild animals, each according to its kind.", "Then the Lord God formed man out of the dust of the earth and He breathed his breath of life into his nostrils and man became a living creature.".

Earth and water—both inorganic—did not produce life *on their own*, but through the action of the Creator's Word. Matter did not become alive randomly; it was animated by the will and power of God.

Even as science breaks down the structure of DNA, catalogues chemical reactions within the cell, and maps the process of cell replication, one profound question remains unanswered: Where did the information come from? We have unlocked many secrets about the composition and function of life, but we still cannot explain how matter came to *know* what to do.

Information is not a material substance. It is conceptual—immaterial—and yet essential. It conveys meaning, coordinates function and enables systems to work in harmony. And it is universally acknowledged that only intelligence produces information.

Throughout this chapter, I have presented concrete and empirical evidence of the vast information embedded in the universe—from DNA to planetary physics. The origin of such intricate and vast information cannot be explained by chemistry alone. Consider this: if the sheer amount of information on the internet overwhelms us, how much more astounding is the information required to build and operate a single human being?

The Christian worldview offers an elegant and consistent explanation: an intelligent Creator designed life with intention and order. In contrast, atheistic theories demand belief in contradictions—violations of physics, biology, and logic—in order to dismiss the need for God.

The academically trained atheist may marvel at the structure of matter, cells, and life, yet often ignores the central issue of information. Take a smartphone: you can disassemble it into its smallest parts and examine every circuit. But without the software—the immaterial code that tells the device what to do—it is merely a paperweight. Applications have no weight or shape, yet they are essential. The same principle applies to life: without information, the cell cannot function.

As a systems engineer, I write applications daily. I must carefully plan and sequence each instruction. If I mistype a command or reverse a step, the system fails. No computer, however powerful, can "figure out" what to do without clear, intelligent input. Life is the same. The genetic code in DNA is not a byproduct of chemistry; it is a structured, symbolic language—an instruction set. Only intelligence writes code.

Darwinian evolution may propose chemical processes to explain the structure of life, but it cannot explain the *origin of information*—which is immaterial, precise, and purposeful. Chemical reactions may assemble molecules, but they do not create syntax, semantics, or symbolic logic. The instructions found in DNA require an author—someone capable not only of generating complex sequences but also of assigning meaning to them. This fundamental gap remains unbridged by purely materialistic explanations and speaks to the necessity of an intelligent cause behind the existence of life.

The invisible presence of information saturating the visible universe is an overwhelming argument against naturalism and materialism—and a resounding affirmation of the biblical Christian worldview. All scientific inquiry uncovers layers of information that point to an intelligent source. As Albert Einstein once said, "Man finds God behind every door that science manages to open." That insight is more relevant today than ever.

If you once believed the improbable because you lacked alternatives, I hope you now find peace in knowing that science and reason do not contradict Scripture—they corroborate it. In fact, the coherence between biblical truth and scientific discovery is more compelling than any naturalistic explanation.

Does God exist? There can be no doubt.

DOES HE COMMUNICATE WITH US?

> You might say to yourself, "How can we know that the Lord did not speak the message?" If what the prophet proclaims in the name of the Lord is not true and it does not happen, then the message was not proclaimed by the Lord. The prophet has spoken presumptuously, you should not fear him.
>
> Deuteronomy 18:21-22

There are many philosophical positions that human beings have taken throughout history concerning the existence and nature of God. These include atheism, agnosticism, anticlericalism, pantheism, panentheism, deism, and theism, among others. Each of these perspectives differs in the degree to which it affirms God's existence and involvement with humanity. While we need not delve into the technical definitions of each, it is worthwhile to explore three of the most significant positions on this spectrum: atheism, deism, and theism.

At one extreme is atheism, which we have already explored at length. In brief, the atheist asserts that God does not exist and that everything in the universe—including life and consciousness—can be explained through natural processes that are observable, measurable, and reproducible within a scientific framework. The atheist places unwavering faith in the power of empirical investigation and regards belief in the divine as a relic of the past.

At the opposite end of the philosophical spectrum is theism. The theist not only believes in the existence of God but also affirms that God is personal, relational, and communicative. For the theist, God is not a distant force but an ever-present Father who actively engages with His creation and reveals Himself through Scripture, providence, prayer, and even miracles. The theist holds that divine interaction is not only possible but essential for understanding the full meaning and purpose of life.

Occupying the middle ground between these two poles is deism. The deist believes in a Creator who brought the universe into existence but subsequently withdrew, leaving creation to operate autonomously according to the laws of nature. In this view, God is like a cosmic watchmaker—He constructed the intricate mechanisms of the universe, set them in motion, and then stepped away. The deist sees no need for ongoing divine intervention and rejects the authority of religious texts, doctrines, or supernatural claims such as miracles and prophecies. Nature itself is seen as the only true "word" of God. Deism is often described as spiritual but not religious—an affirmation of divine origin without divine presence.

So, what causes someone to choose deism over theism? In general, there are three major reasons:

- ***The Problem of Evil and Suffering***: Many struggles to reconcile the existence of suffering, injustice, and evil with the notion of a loving and omnipotent God. If God is good and powerful, why does He allow such pain to exist?
- ***Religious Pluralism***: The sheer number of religions—each claiming exclusive access to divine truth—leads some to skepticism. How can one know which, if any, is true?
- ***Scientific Explanations***: As science continues to explain natural phenomena that were once considered divine mysteries, some conclude that the supernatural is unnecessary. They accept that a Creator exists but believe He has no ongoing role in the affairs of the world.

These concerns are not trivial, and they deserve thoughtful, respectful responses. However, when approached with humility, honesty, and an openness of heart, it becomes clear that the presence of God is not distant or theoretical but profoundly personal and enduring.

Through reflection on one's life, many begin to recognize the signs of divine love, guidance, and care—often subtle, but undeniably real.

It is tragic when a person acknowledges the evidence of Creation and affirms God as Creator yet stops short of knowing Him as Father. To believe in a God who creates but does not care is to miss the deepest truth of our existence: that we are *children* of God. That relationship—not mere acknowledgment—offers the highest joy and meaning in life.

Is not the universe itself a testimony to God's intelligence and power? Does not the beauty, order, and complexity of nature compel us to consider a divine mind behind it all? And if God is indeed so wise and powerful, could He really have created all this and then chosen silence? Could such a Creator have refused to build communication bridges with His most cherished creation—human beings? Could He be an extraordinary architect, yet an absent and indifferent Father?

Of course not.

God has not remained distant or silent in His relationship with humanity. Throughout history, He has chosen to communicate with us in four fundamental ways, each revealing different dimensions of His nature and His desire for connection with His creation.

- **Through Natural Revelation**. The natural world speaks continually of its Creator. As the psalmist beautifully declares: "The heavens proclaim the glory of God; the firmament shows forth the work of his hands. One day imparts that message to the next, and night conveys that knowledge to night. All this occurs without speech or utterance; no voice can be heard." (Psalms 19:1–4). Since the earliest days of human existence, even in primitive societies, people have looked at the stars, the seas, the mountains, and the miracle of life itself and intuitively recognized the presence of a transcendent, all-powerful Being. Through creation, they have discerned God's perfection, generosity, creativity, patience, tenderness—and even His joy and sense of humor. The natural world, in all its beauty and order, has always served as a silent yet unmistakable testament to His existence and nature.

- **Through Evangelical Revelation**. Scripture makes it clear that God has chosen to speak directly to humanity through His prophets. Phrases such as "All this took place in order to fulfill what the Lord had announced through the prophet" (Matthew 1:22), and "God fulfilled what He had foretold through all the Prophets, revealing that his Christ would suffer" (Acts 3:18), affirm that the prophetic voice is God's chosen instrument to convey His will. Prophets were not self-appointed; they were selected by God to speak on His behalf. As Amos writes, "Indeed, the Lord God does nothing without revealing his plan to his servants, the prophets." (Amos 3:7). The story of Moses illustrates this relational dynamic vividly. When Moses hesitated to speak to Pharaoh due to his stammer, God said to him: "Do you not have a brother, Aaron, a Levite? I know that he can speak well... You will speak to him and place the words he is to say in his mouth. I will be with you and with him while you speak... He will speak to the people for you. It will be as if he is your mouth and you are his God." (Exodus 4:14–16). This passage reflects the intimacy, clarity, and authority with which God entrusted His message to human agents.

- **Through Jesus of Nazareth**. The most complete and personal communication of God is found in His Son. "In previous times, God spoke to our ancestors in many and various ways through the prophets, but in these last days He has spoken to us through His Son, whom He appointed heir of all things and through whom He created the universe." (Hebrews 1:1–2). In Jesus, the invisible God became visible. He lived among us, not merely to teach, but to embody truth, grace, and love in human form. Through Jesus, God's voice is unmistakable—full of mercy, justice, and authority. Christ's words, actions, and sacrifice resolved all ambiguities and fulfilled the message proclaimed by the prophets. In Jesus, God not only spoke to us but walked with us.

- **Through Our Feelings and Personal Experiences**. God continues to communicate today in the most intimate language of all: the language of the heart. Saint Ignatius of Loyola once paused in a garden, gazed long at a single flower, then gently tapped it with his cane, and said, "Stop shouting that God loves me!" In that moment of silent wonder, the truth of divine love overwhelmed him. Just as couples who deeply love one another can communicate without words—through a look, a gesture, a shared silence—so too does God speak to us through the movements of the soul. Emotions such as joy, serenity, empathy, charity, hope, awe, compassion, justice, patience, and peace are all channels through which God's presence is revealed. These experiences are not mere sentiments; they are the spiritual resonance of a relationship between Creator and creature. God speaks continuously in the quiet whispers of our inner life.

Among these four modes of divine communication, two are uniquely preserved and proclaimed in the Bible: The Evangelical Revelation and the Revelation through Jesus Christ. For this reason, we can rightly affirm that the Bible is the *Word of God*. Though written by human authors with distinct voices, languages, and literary styles, they were chosen and inspired by God to transmit His truth to every generation. In it, God's voice is not only heard—it is safeguarded, remembered, and made alive for those who seek Him.

The *Dogmatic Constitution on Divine Revelation, Dei Verbum*, promulgated by the Second Vatican Council, offers profound insight into the mystery of how God chose to reveal Himself to humanity. It states:

> In composing the sacred books, God chose men and while employed by Him they made use of their powers and abilities, so that with Him acting in them and through them, they, as true authors, consigned to writing everything and only those things which He wanted. (*Dei Verbum*, 11)

The Bible did not descend from heaven as a finished product. It was neither composed and sealed in heaven nor delivered by an angel to a pastor or to the ruling empire of the time for divine authentication. Rather, it was written—under divine inspiration—by very human individuals, chosen for a sacred task. These authors, though guided by God, were fully human, with personalities, cultures, limitations, and perspectives not unlike our own.

The same constitution continues:

> Therefore, since everything asserted by the inspired authors or sacred writers must be held to be asserted by the Holy Spirit, it follows that the books of Scripture must be acknowledged as teaching solidly, faithfully and without error that truth which God wanted put into sacred writings for the sake of salvation. (*Dei Verbum*, 11)

Two phrases stand out here: "without error" and "for the sake of salvation." From the standpoint of modern knowledge in the 21st century, we may encounter details in Scripture that appear inconsistent with geography, history, chronology, or science. But these do not constitute errors in what matters most. The Bible is not a science textbook or a historical ledger; it is a divinely inspired revelation oriented toward a singular purpose: our eternal salvation. In this sense, it is inerrant—it teaches truth without error where that truth concerns God's saving will.

As Saint Paul writes:

> All Scripture is inspired by God and is useful for teaching, for refutation, for correction, and for training in uprightness, so that the man of God may be proficient and equipped for good work of every kind. (2 Timothy 3:16-17)

Interestingly, the word Bible itself does not appear within the Bible. Scripture refers to itself as "the Word of God" or simply "Scripture." The term Bible comes from the Greek word "biblion", meaning "scroll" or "book." The plural form Bible— "the books"—was later adopted into Latin as a singular feminine noun, thus referring to the Bible as the Book par excellence.

When we affirm that the Bible is the Word of God, we do not mean that it is merely a "word" in the linguistic sense—a phonological unit found in a dictionary. Rather, we are referring to something far more profound. Though this Word is human—written by and for human beings—it is also divine in its origin and authority. It bridges heaven and earth, time, and eternity.

We speak to God through prayer. He speaks to us through His Word.

ARGUMENT: THE HOLY SPIRIT IS THE AUTHOR OF THE BIBLE

Some Christians find the Gospel of the Beloved Disciple—commonly attributed to John—"difficult" to read because of its profound theological depth and contemplative tone, which distinguish it from the other three Gospels. Yet in its closing verse, it offers a statement that is universally accessible and deeply moving:

> But there are also many other things that Jesus did; and if every one of them was recorded, I do not think the world itself could contain the books that would be written (John 21:25)

This powerful declaration reminds us that the Gospels present only a portion of what Jesus said and did. The apostles, having spent years in the intimate presence of the Master, were recipients of countless teachings that were never written down. Yet these unwritten words and moments were by no means insignificant. One can imagine the many tender and instructive encounters they shared—gathered around a fire by the Sea of Tiberias, enjoying freshly caught fish and good wine as Jesus unfolded parables, answered questions, and patiently reshaped their minds and hearts.

To form His followers for their mission, Jesus had to teach them how to think, how to live, and how to love—again and again, in every conceivable circumstance. Before His ascension, He entrusted them with a commission that extended to all nations: *"Teach them to observe all that I have commanded you."* (Matthew 28:20) The apostles fulfilled this charge faithfully throughout their lives, and before their earthly journey ended, they ensured that Jesus' teachings would endure. They passed them on—faithfully and fully—to their successors, who passed them to theirs, in an unbroken line that reaches today's bishops. This living transmission, animated and safeguarded by the Holy Spirit, is what the Church calls Sacred Tradition.

Alongside Sacred Scripture, this Sacred Tradition forms the dual source of Divine Revelation—one oral, one written. Both are sacred, and both are necessary. Scripture and Tradition together preserve the fullness of God's self-revelation. For example, the written Word does not explicitly tell us what happened to the Virgin Mary at the end of her earthly life. Yet from the earliest centuries, the Church—through Tradition—has consistently proclaimed her Assumption into heavenly glory. Cities founded many centuries ago, such as those named *Asunción*[72], bear witness to this cherished belief. It was not until November 1, 1950, that this belief was formally defined as dogma by Pope Pius XII. Sadly, when many Protestant communities chose to reject Tradition, they unwittingly severed themselves from this immense wellspring of spiritual knowledge and continuity.

Earlier in this chapter, it was noted that the Bible contains two of the four primary ways in which God communicates with us: the Evangelical Revelation (through the words of the prophets), and the Revelation through the life, ministry, and teachings of Jesus Christ. In the first case, prophets spoke not by their own will, but as instruments of God, using their own voices, cultures, and styles to transmit the divine message.

According to *Merriam-Webster*, prophecy is defined as *"the inspired declaration of divine will and purpose, a prediction of something to come."* By contrast, a *prediction* is simply *"a declaration that something will happen in the future."* The difference lies in the source: prophecy presumes divine origin, while prediction may stem from observation, intuition, or guesswork. In the same way, the term *prophet* is defined as *"a person who possesses the gift of prophecy,"* whereas a *fortuneteller* is *"someone with the supposed ability to foretell future events, especially specific personal outcomes."* Again, the essential distinction is one of divine authority versus human speculation.

---

[72]Asunción is the capital city of Paraguay, founded on August 15, 1537.

Nueva Guatemala de la Asunción is the capital of the Republica the Guatemala, founded on January 23, 1776.

Nuestra Señora de la Asunción de Panamá was the original name of Ciudad de Panamá, the capital of Panamá, founded on August 15, 1519.

Throughout history, both prophets and fortune tellers have made claims about the future—some eerily accurate, others wildly off the mark. Their reputations rise or fall depending on whether their statements are fulfilled. The problem, however, is that many of these so-called prophecies are written in language so vague, cryptic, or ambiguous that any event could be interpreted as a fulfillment. This ambiguity allows for retrospective interpretation, giving the illusion of foresight where none existed.

Beginning in 2012, so-called "Mayan prophecies" began circulating widely on social media and soon spread through mass media outlets. One prophecy in particular captured public imagination—it predicted the end of the world on the winter solstice of that year. The buzz reached such heights that Hollywood produced a blockbuster film titled 2012, directed by Roland Emmerich, which dramatized these apocalyptic predictions. The film was seen by over 140 million viewers in North America alone and offered a cinematic interpretation of what the supposed prophecy foretold.

The source of this doomsday scenario was *Chilam Balam of Chumayel*, a collection of 16th- and 17th-century writings composed in the Yucatán Peninsula in the Mayan language. These books contain accounts of historical and prophetic events relating to Mayan civilization. The name "Chumayel[73]" refers to the town where the manuscripts originated. In the 1933 English translation by Ralph L. Roys, one prophecy—cited as the inspiration for the 2012 theory—reads:

> On the thirteenth Ahau at the end of the last katun, the Itza will be rolled and the Tanka will roll, there will be a time in which they will be submerged in darkness and then the men of the sun will come bringing the future sign; the land will wake up in the north, and in the west, the itza will wake up.

How this poetic and ambiguous passage came to be interpreted as predicting the end of humanity on a specific date is difficult to comprehend.

---

[73]Chumayel is a town in the state of Yucatán, Mexico, and serves as the administrative center of the municipality of the same name. It is located approximately 70 kilometers southeast of Mérida, the state capital, and about 20 kilometers north of the city of Tekax.

The same tendency to retroactively impose meaning onto cryptic texts is seen in the case of Michel de Nostredame, better known as Nostradamus. A 16th-century French physician and astrologer, he published *Les Prophéties* in 1555—a collection of 942 poetic quatrains that purported to foretell future events. His method of divination involved a reflective ritual using a brass tripod and a bowl of water, through which prophetic visions would come to him in the form of flames or images.

Despite their opaque content, his writings found a receptive audience, including royalty. Catherine de' Medici, Queen of France, was so impressed by his prediction of King Henry II's death in a jousting tournament that she summoned Nostradamus to court, honored him lavishly, and housed him in her palace.

Nostradamus became a celebrity, and people from across France sought his insight. To keep up with demand, he began composing what he called Centuries—prophetic verses grouped in sets of one hundred. Anticipating scrutiny from the Inquisition, he deliberately obscured his messages by using riddles and mixing several languages, including Latin, Greek, Provençal, Italian, Hebrew, and Arabic.

Supporters argue that Nostradamus's prophecies have proven accurate—although their interpretations are invariably offered after the events in question. Take, for instance, the aftermath of Princess Diana's death in 1997. Following the tragedy, Nostradamus's followers pointed to Quatrain XXVIII:

> The penultimate son of the man with the Prophet's name, will bring Diana to her day of rest; He will wander far because of a frantic head, delivering a great people from subjection.

Similarly, after the September 11, 2001, terrorist attacks, another verse was widely circulated online:

> Two steel birds will fall from the sky on the Metropolis. The sky will burn at forty-five degrees latitude. Fire approaches the great new city. Immediately a huge, scattered flame leaps up. Within months, rivers will flow with blood. The undead will roam the earth for little time.

And concerning the atomic bombings of Hiroshima and Nagasaki in August 1945, interpreters cited:

> Near the gates and within two cities there will be scourges the like of which was never seen: famine within plague, people put out by steel, crying to the great immortal God for relief.

But how do these verses, steeped in symbolic language and open to multiple interpretations, truly constitute prophecies? Most claims of fulfillment arise after the fact, often forcing events to fit the text through selective interpretation and creative reconstruction.

Now contrast this with a biblical prophecy, one spoken by Jesus Himself, recorded plainly and directly:

> As Jesus left the temple and was walking away, his disciples came up to him to call his attention to the buildings of the temple. Then He said to them, "Do you see all these? Amen, I say to you, not one stone here will be left upon another; every one will be thrown down." (Matthew 24:1-2)

This prophecy was fulfilled with chilling accuracy in the year AD 70, during the First Jewish–Roman War. The Roman army, led by Titus[74], the future emperor, and supported by Tiberius Julius Alexander, lay siege to Jerusalem, breaching its walls and demolishing its most sacred structures.

Flavius Josephus, a Jewish historian who was both a witness to and participant in the events, documented the horrifying aftermath. He wrote:

> Now, as soon as the army had no more people to slay or to plunder, because there remained none to be the objects of their fury (for they would not have spared any, had there remained any other such work to be done) Caesar gave orders that they should now demolish the entire city and temple, but should leave as many of the towers standing as were of the greatest eminency;

---

[74]Titus Flavius Sabinus Vespasianus, commonly known as Titus (December 30, 39 – September 13, 81), was Roman Emperor from AD 79 until his death in AD 81. He was the second emperor of the Flavian dynasty, which ruled the Roman Empire from 69 to 96. This lineage includes the reigns of his father, Vespasian (69–79), his own (79–81), and that of his brother, Domitian (81–96).

that is, Phasaelus, and Hippicus, and Mariamne; and so much of the wall as enclosed the city on the west side. This wall was spared, in order to afford a camp for such as were to lie in garrison, as were the towers also spared, in order to demonstrate to posterity what kind of city it was, and how well fortified, which the Roman valor had subdued; but for all the rest of the wall, it was so thoroughly laid even with the ground by those that dug it up to the foundation, that there was left nothing to make those that came thither believe it had ever been inhabited. This was the end which Jerusalem came to by the madness of those that were for innovations, a city otherwise of great magnificence, and of mighty fame among all humanity.

Ain-Karim, a tiny town situated seven kilometers west of Jerusalem in the Judean region, was the site of a prophecy at the dawn of the Christian era. Mary, who was pregnant and eagerly anticipating the birth of Jesus, visited Elizabeth, who lived there with her spouse Zechariah. The Virgin Mary sang the *Magnificat*, a hymn of praise to God, following the initial greeting. She prophesied that "All generations will call me blessed." (Luke 1:48).

Imagine this moment: a girl perhaps fifteen years old, poor, unknown, unmarried, and from a tiny village, proclaims with absolute confidence that every generation of humanity will call her blessed. And now, over two thousand years later, Mary's name is indeed honored across cultures, languages, and continents. Her image graces churches, her name is echoed in liturgy, and her role in salvation history is universally recognized among Christians. Against all odds, this prophecy, too, has come true.

Despite being composed over 1,700 years, by more than forty different authors, from diverse walks of life, in varied locations, and across multiple eras, the books of the Bible stand in astonishing harmony. They speak with a unified voice about God's character, human nature, the path to salvation, and the promise of eternal life. The consistency of its message across such a vast span of time and geography defies human explanation.

Over the past century, public health efforts have shifted their focus— from combating problems related to malnutrition to addressing the growing crisis of obesity. The rise of industrialization has impacted every aspect of human productivity, including agriculture. In response to escalating demand, the food industry adopted increasingly efficient methods of mass production.

Take the *Gallus Bankiva*, a breed of chicken domesticated around nine thousand years ago, originally bred to produce just one egg per month. Today, commercial poultry farmers raise *New Hampshire* and *Leghorn* varieties, which can lay up to three hundred eggs per year. This drive toward large-scale, cost-effective output has flooded supermarket shelves with processed, calorie-dense foods. Unsurprisingly, just a few generations later, populations began to experience noticeable weight gain.

In response, diets emerged—quickly evolve into a booming industry. The market is now saturated with weight-loss solutions, each claiming superior effectiveness. Some extreme plans promise 10% weight loss in just two weeks, while others make similar claims without requiring any exercise. Certain diets tout the benefits of increased fat intake, while others demand an almost total elimination of fat. Some discourage vegetables in the early stages; others recommend incorporating them immediately. Dairy is eliminated by some, while others allow it. Fruit is forbidden by a few diets—at least temporarily—and alcohol is banned outright in some regimens but permitted in moderation by others. And yet, many of these trendy practices come with serious health warnings.

A century ago, the public had no access to literature on dieting. Today, bookstores have entire sections dedicated to weight loss and nutrition, filled with works written by medical professionals—physicians, dietitians, endocrinologists, and scientists. And yet, with so many books to choose from, it is impossible to find two that do not directly contradict each other.

Now, I invite the reader to consider this challenge: Choose any library in the world and try to assemble seventy-three books written over a span of seventeen hundred years by at least fifty different authors, from different cultures and continents, which contain 2,500 prophecies—95% of which have already been fulfilled—and that consistently address three major themes, all without a single contradiction.

The Holy Bible is composed of seventy-three books, written over a span of one thousand seven hundred years by at least fifty different authors who lived on three different continents. Despite this extraordinary diversity in time, geography, and background, the Bible revolves around three unified themes: salvation, the Church, and the Kingdom of Heaven.

It contains more than 2,500 prophecies, of which over 2,380 have already been fulfilled—and these fulfillments can be historically verified. Even more remarkably, there is no contradiction among these core themes. From the first verse of Genesis to the final word of Revelation, there is an unmistakable thread of coherence and consistency throughout.

Such unity is humanly impossible across so many centuries and by authors. The only way this could occur is if the entire Bible has one ultimate Author: The Holy Spirit. God revealed His will through these chosen individuals, guiding them to record exactly what He intended for humanity to know.

God's unique method of communication with us unfolds through two channels: oral and written revelation. Through them, He invites us to understand His truth, to trust in His plan, and to enter into a relationship with Him—across generations, cultures, and ages.

FIRST THESIS: HISTORICAL SUPPORT FOR THE BIBLE

The question of the "originals" of the Holy Scriptures frequently arises during my lectures on biblical topics.

To illustrate the concept of an "original" document, consider the manuscript of the *United States Declaration of Independence*, signed by John Hancock. This historical document is currently housed in the National Archives[75], specifically in the Rotunda for the Charters of Freedom. Adjacent to Hancock's signature are those of future presidents Thomas Jefferson and John Adams, along with other prominent figures such as Benjamin Franklin—individuals whose names have been etched into history through the events surrounding the Declaration. Despite the passage of time since July 4, 1776, when they signed the document, their signatures remain faintly visible. This is due to the rudimentary conservation techniques available at the time.

This document serves as a clear example of what we can rightfully call an "original." Why? Because its chain of custody has been meticulously preserved and documented. A comprehensive record exists detailing the document's origin, previous custodians, storage conditions, restorations, and more. This rigorous documentation allows us to confidently refer to it as the "original" Declaration of Independence.

In contrast, while biblical manuscripts are of immense historical and theological significance, we lack the kind of uninterrupted chain of custody that would allow us to assert, for example, that a particular scroll is the original Book of Genesis written by Moses. This uncertainty is due to several factors. First, the common materials used by ancient scribes— parchment and vellum—are highly susceptible to deterioration from light, moisture, and handling, unless carefully preserved. Second, we do not possess a verified handwriting sample of Moses to compare with any manuscript. These issues are not unique to biblical texts; they affect all literary works from antiquity. For instance, how do we verify the authenticity of a manuscript of Homer's *Iliad*? Even if a papyrus were

---

[75]The National Archives and Records Administration (NARA) is an independent agency of the United States federal government responsible for preserving and documenting government and historical records.

found and accurately dated to the time Homer is believed to have lived, we would still need to demonstrate that the handwriting is genuinely his.

Does this mean the Bible is discredited because we do not possess its original manuscripts? Absolutely not. If that were the standard, then the entirety of human literary heritage—spanning more than five millennia— would also be invalidated.

In *An Introduction to Research in English Literary History*, Chauncey Sanders, a respected authority in the field of documentary research, outlines three foundational principles of historiography[76] and paleography[77] that are crucial for evaluating ancient texts:

- **The Bibliographic Test**: This test assesses the reliability of copies of an ancient document. It compares various reproductions, often across different languages, to establish textual accuracy. While God inspired the authors of Scripture, the process of copying was not divinely protected from human error. The more copies available, the better, as this enables comparative analysis to determine textual lineage and proximity to the original. The closer a manuscript is in date to the time of its original composition, the greater its evidentiary value.
- **The Internal Test**: This test seeks to identify and understand discrepancies among copies. It evaluates whether variations stem from unintentional errors, deliberate alterations, or natural developments in language and grammar over time.
- **The Test of External Evidence**: This principle involves corroborating a document's content through other independent sources or archaeological findings. Such external validation helps verify the historical reliability of the events or details described in the manuscript under examination.

I will apply these three critical tests—bibliographic, internal, and external—first to the New Testament and then to the Old Testament, to

---

[76]Science that studies history.

[77]Paleography is the science responsible for deciphering ancient writings and studying their evolution, in addition to dating, locating, and classifying the various written records it examines.

demonstrate that the Bible, despite its antiquity, faithfully preserves the words originally written by the prophets. There are only two notable exceptions: first, the Bible we read today has been translated into English; and second, it is rendered in contemporary language to ensure clarity and accessibility.

To illustrate how language evolves over time, consider *The Canterbury Tales*, the magnum opus of the English poet Geoffrey Chaucer[78]. First composed around 1400, this work, when compared with a modern edition, reveals striking changes in vocabulary, spelling, grammar, and syntax[79].

Original English	Current English
Whan that Aprill with his shoures soote The droghte of March hath perced to the roote, And bathed every veyne in swich licour Of which vertu engendred is the flour; Whan Zephirus eek with his sweete breeth Inspired hath in every holt and heeth The tendre croppes, and the yonge sonne Hath in the Ram his half cours yronne, And smale foweles maken melodye, That slepen al the nyght with open ye (So priketh hem Nature in hir corages), Thanne longen folk to goon on pilgrimages, And palmeres for to seken straunge strondes, To ferne halwes, kowthe in sondry londes; And specially from every shires ende Of Engelond to Caunterbury they wende, The hooly blisful martir for to seke, That hem hath holpen whan that they were seeke.	When April with its sweet-smelling showers Has pierced the drought of March to the root, And bathed every vein (of the plants) in such liquid By which power the flower is created; When the West Wind also with its sweet breath, In every wood and field has breathed life into The tender new leaves, and the young sun Has run half its course in Aries, And small fowls make melody, Those that sleep all the night with open eyes (So Nature incites them in their hearts), Then folk long to go on pilgrimages, And professional pilgrims to seek foreign shores, To distant shrines, known in various lands; And specially from every shire's end Of England to Canterbury they travel, To seek the holy blessed martyr, Who helped them when they were sick.

---

[78]Geoffrey Chaucer (1340s – October 25, 1400) was an English poet, author, and civil servant, best known for *The Canterbury Tales*. He is often referred to as the "father of English literature" or, alternatively, the "father of English poetry."

[79]Translator's Note: The original version used *El Ingenioso Hidalgo Don Quijote de la Mancha* as an example to illustrate the evolution of language.

The bibliographical proof of the New Testament. In *Harvest of Hellenism: A History of the Near East from Alexander the Great to the Triumph of Christianity*, Francis Edward Peters, Emeritus Professor of History at New York University (NYU), affirms:

> Based on the manuscript tradition alone, the works that made up the Christian's New Testament were the most frequently copied and widely circulated books of antiquity.

This statement underscores the exceptional preservation of the New Testament. Its authenticity is rooted in the extraordinary number of manuscript copies that serve as textual witnesses, attesting to the original sources. These manuscripts, preserved over centuries, reflect the careful transmission of the biblical text.

There are over 5,686 Greek manuscripts—either complete or partial—of the New Testament. These were hand-copied from the late first century through the invention of the printing press in the 15th century. The abundance of these manuscripts allows scholars to compare versions, trace textual variants, and identify the most accurate representations of the original texts.

Beginning in the third century, the New Testament was translated into several major languages of the ancient world: Coptic, Syriac, and Latin.

Among these, Latin had the greatest influence in the Western world. The Latin *Vulgate*[80], translated by Saint Jerome[81] in 382, became the authoritative version for many centuries[82]. Today, there are more than 10,000 surviving manuscripts of the *Vulgate*.

---

[80]The *Vulgate* is a translation of the Hebrew and Greek Bible into Latin.

[81]Eusebius Hieronymus (c. 340, Stridon, Dalmatia – September 30, 420, Bethlehem), commonly known as Saint Jerome—also referred to as Jerome of Stridon or simply Jerome—was a Christian scholar and theologian. At the request of Pope Damasus I, he translated the Bible into Latin, a version later known as the Vulgate. Pope Damasus had previously convened the Council of Rome in 382 to establish the first canon of biblical books. Saint Jerome is recognized as a Father of the Church and is one of the four great Latin Fathers.

[82]The first complete Bible in English was published abroad—most likely in Antwerp— in 1535. It was translated by Myles Coverdale (1488–1569), an Augustinian friar from Yorkshire educated at Cambridge, who claimed to have "faithfully and truly translated

In total, including manuscripts in other languages, there are over 20,000 known manuscript copies of the New Testament—either in full or in part. This unparalleled abundance strengthens confidence in the text's authenticity and provides a robust foundation for textual comparison and historical verification.

To appreciate the manuscript wealth of the New Testament, consider Homer's *Iliad*—the most widely known and copied classical Greek text. There are only 643 surviving manuscript copies of the *Iliad*. The earliest fragment, dating to approximately AD 150, consists of 16 Greek pages and is currently displayed in the British Library. The earliest complete manuscript of the *Iliad* does not appear until the 13th century, over 2,000 years after its original composition.

This comparison highlights the unique position of the New Testament in textual history: no other ancient document is as comprehensive and closely preserved.

Another important early version of the Bible is the *Syriac Peshitta*, a translation directly from the Hebrew scriptures. Produced around the second century, the *Peshitta* played a significant role in the transmission of the biblical text in the East. Over 350 manuscript copies of the *Peshitta*, dating from the fifth century onward, still exist today.

The accompanying chart (with approximate dates and ages) describes in greater detail the fate of various ancient works, including the Old and New Testament manuscripts:

Author: Book	Year of writing	Older copy	Difference in years	Number of copies
Homer: *The Iliad*	BC 800	BC 400	400	643
Julius Caesar: *Commentary on the Gallic Wars*	BC 100	AD 900	1000	10

[it] out of Douche [German] and Latin into English." Working independently, Coverdale revised William Tyndale's New Testament and drew upon several sources for his translation, including Martin Luther's German Bible, the Zürich Bible, the Latin Vulgate, and another Latin translation of the Hebrew Old Testament.

Tacitus: The *Annals*	AD 100	AD 1100	1000	20
Pliny the Elder: *Natural History*	AD 100	AD 850	750	7
Plato: *Dialogues*	BC 400	AD 900	1300	7
Thucydides: *History of the Peloponnesian War*	BC 460	AD 900	1300	8
Old Testament	BC 1445-135	BC 625 (fragment) BC 135 (almost the entire OT)	820-0	5,686 (2,600,000 pages in total) in original language
New Testament	AD 50-100	AD 114 (fragment) AD 200 (books) AD 250 (almost all NT) AD 325 (all the NT)	39 100 150 225	45,000 in other languages

It is important to recognize that not all ancient documents carry the same value or weight. A small fragment cannot be placed on the same level as a complete manuscript. The age and significance of such documents are determined through various factors, including the color and texture of the ink and parchment, the style and shape of the letters, the presence of ornamentation, the use of punctuation, textual divisions, and the materials employed in their creation. These physical and stylistic features help scholars establish not only the approximate age of a manuscript but also its origin and reliability. With these criteria in mind, we can examine some of the most notable biblical manuscripts based on their antiquity, physical condition, and degree of completeness at the time they were discovered.

- The *Rylands Library Papyrus*, also known as P52 or "The Fragment of Saint John," is the earliest known New Testament manuscript. Dated to around AD 125 and preserved at the John Rylands Library in Manchester, this small papyrus fragment contains a portion of the Gospel of John—verses 18:31–33 on the front and 18:37–38 on the back. Despite its brevity, its early date makes it a crucial witness to the existence and circulation of the Gospel narrative within decades of its composition.

- The *Codex Sinaiticus*, discovered by Constantin von Tischendorf in 1844 at the Monastery of Saint Catherine at Mount Sinai, dates to approximately AD 350. Written in Greek, this codex includes nearly all the New Testament along with a significant portion of the Old Testament (in its Septuagint form). It is now housed in the British Library in London and remains one of the most complete and valuable biblical manuscripts from antiquity. It is also available for public viewing online at www.codexsinaiticus.org.

- The *Codex Vaticanus*, preserved in the Vatican Library and documented as early as 1475, is another foundational biblical manuscript from the fourth century. Written in Greek, it contains nearly the entire Old Testament and the majority of the New Testament. Its early dating, exceptional quality, and textual consistency make it a cornerstone document in the study of biblical transmission.

- The *Codex Alexandrinus*, dating to the fifth century, was presented to King Charles I of England by the Patriarch of Constantinople in 1627 and is now located in the British Library in London. Written in Greek, it contains an almost complete version of the Old Testament (Septuagint) and the full New Testament. Of the three great codices, *Alexandrinus* is the most complete and remains a vital reference point for biblical scholars analyzing early Christian texts.

The test of the New Testament's external evidence. Bishop Eusebius of Caesarea is often regarded as the "Father of Church History" for his foundational work in documenting the early centuries of Christianity. His Ecclesiastical History, likely written in the early third century, provides the earliest comprehensive account of the development of the Christian Church. In this work, Eusebius references several writings from earlier Church figures, including letters by Bishop Papias of Hierapolis, an Apostolic Father whose works are dated to around AD 130. He wrote:

Mark, having been the interpreter of Peter, wrote accurately, though not in order, all that he remembered of the things said or done by the Lord. For he had neither heard the Lord nor been his follower, but afterward, as I said, he was the follower of Peter, who gave his instructions as circumstances demanded, but not as one giving an orderly account of the words of the Lord. So that Mark was not at fault in writing certain things as he remembered them. For he was concerned with only one thing, not to omit anything of the things he had heard, and not to record any untruth in regard to them. (Book III: XXXIX ,15)

Irenaeus of Lyons, known as Saint Irenaeus, wrote:

For, since there are four zones of the world in which we live, and four principal winds ... the Artificer of all ... has given us the Gospel under four aspects but bound together by one Spirit.
Matthew also issued a written Gospel among the Hebrews in their own dialect, while Peter and Paul were preaching at Rome, and laying the foundations of the Church. After their departure, Mark, the disciple and interpreter of Peter, also handed down to us in writing what had been preached by Peter. Luke also, the companion of Paul, recorded in a book the Gospel preached by him. Afterwards, John, the disciple of the Lord, who also had leaned upon His breast, did himself publish a Gospel during his residence at Ephesus in Asia. (Against Heresies Bk. 3.11.8)

Additional external sources worth consulting are the historical accounts written by contemporaries of Jesus.

Cornelius Tacitus, born around AD 55 in Gallia Narbonensis—a Roman province at the time—rose to become one of the most prominent historians of the Roman Empire. He served as both consul and provincial governor, earning a reputation for his sharp political insight and literary precision. Among his various works, his most notable contributions are *the Annals* and the *Histories*, which chronicle the reigns of Roman emperors and significant events of the early imperial period. In the *Annals*, Tacitus makes a noteworthy declaration:

Consequently, to get rid of the report, Nero fastened the guilt and inflicted the most exquisite tortures on a class hated for their abominations, called Christians by the populace. Christus, from whom the name had its origin, suffered the extreme penalty during the reign of Tiberius at the hands of one of our procurators, Pontius Pilatus, and a most mischievous superstition, thus checked for the moment, again broke out not only in Judea, the first source of the evil, but even in Rome, where all things hideous and shameful from every part of the world find their center and become popular. (Book 15,44)

The "most mischievous superstition" is a possible reference to the resurrection of Jesus.

Gaius Suetonius Tranquillus, commonly known as Suetonius, was a Roman historian and biographer active during the reigns of Emperors Trajan and Hadrian. A member of Pliny the Younger's intellectual circle, Suetonius, later served in Hadrian's imperial court until a series of disagreements led to his dismissal. His most significant work, *De Vita Caesarum* (The Lives of the Caesars), chronicles the lives of the Roman emperors from Julius Caesar to Domitian. In his biography of Emperor Claudius, Suetonius confirms an event also recorded in Acts 18:2: "Since the Jews constantly made disturbances at the instigation of Chrestus, he [Claudius] expelled them from Rome." This likely refers to early conflicts between Jewish communities and followers of Christ (Chrestus being a variant of Christus). In his account of Emperor Nero, Suetonius references the persecution that followed the Great Fire of Rome: "Punishment was inflicted on the Christians, a class of people given to a new and mischievous superstition." The phrase "new and mischievous superstition" is widely understood to refer to belief in Jesus' resurrection—a core tenet of early Christianity.

Flavius Josephus, born in AD 37 in Jerusalem as Joseph ben Matityahu, was a Jewish historian of priestly and royal Hasmonean descent. A highly educated and prolific writer, he composed *Jewish Antiquities* in Greek between AD 93 and 94. This twenty-volume work sought to present a complete history of the Jewish people, from Creation to the outbreak of the Jewish revolt against Rome in AD 66. Among his many references to figures and events mentioned in the New Testament, three are especially significant. One such reference concerns James the Just, the son of Alphaeus[83] and author of the New Testament epistle bearing his name (not to be confused with James, the son of Zebedee).

---

[83]First bishop of Jerusalem, stoned to death in AD 62.

Josephus mentions James in a passage that highlights his role as a key figure in the early Christian community. He wrote:

> Ananus was of this disposition, he thought he had now a proper opportunity [to exercise his authority]. Festus was now dead, and Albinus was but upon the road; so he assembled the Sanhedrim of judges, and brought before them the brother of Jesus, who was called Christ, whose name was James, and some others, [or, some of his companions]; and when he had formed an accusation against them as breakers of the law, he delivered them to be stoned. (Book 20,9)

The second mention is about John the Baptist:

> Now some of the Jews thought that the destruction of Herod's army came from God, and that very justly, as a punishment of what he did against John, that was called the Baptist: for Herod slew him, who was a good man, and commanded the Jews to exercise virtue, both as to righteousness towards one another, and piety towards God, and so to come to baptism .... Now when [many] others came in crowds about him, for they were very greatly moved [or pleased] by hearing his words, Herod, who feared lest the great influence John had over the people might put it into his power and inclination to raise a rebellion, (for they seemed ready to do anything he should advise,) thought it best, by putting him to death, to prevent any mischief he might cause, and not bring himself into difficulties, by sparing a man who might make him repent of it when it would be too late. Accordingly, he was sent a prisoner, out of Herod's suspicious temper, to Macherus, the castle I before mentioned, and was there put to death. (Book 18,5)

And the last mention is about Jesus himself:

> Now there was about this time Jesus, a wise man, if it be lawful to call him a man; for he was a doer of wonderful works, a teacher of such men as receive the truth with pleasure. He drew over to him both many of the Jews and many of the Gentiles. He was [the] Christ. And when Pilate, at the suggestion of the principal men amongst us, had condemned him to the cross, those that loved him at the first did not forsake him; for he appeared to them alive again the third day; as the divine prophets had foretold these and ten thousand other wonderful things concerning him. And the tribe of Christians, so named from him, are not extinct at this day. (Book 18,3)

According to a study published in *International Geology Review*, Volume 54, Issue 15 (2012[84]), geologist Jefferson Williams of *Supersonic Geophysical*, along with colleagues Markus Schwab and Achim Brauer from the *German Research Center for Geosciences*, conducted an in-depth analysis of the subsoil beneath the beach at Ein Gedi, located on the western shore of the Dead Sea. Their research uncovered deformed sediment layers, which provide geological evidence of at least two significant seismic events that impacted the region. The first was an earthquake dated to BC 31, while the second occurred sometime between AD 26 and 36. Notably, this second event aligns chronologically with the earthquake described in Matthew 27 of the New Testament, which is said to have occurred at the moment of Jesus' crucifixion.

The bibliographic proof of the Old Testament. The number of surviving Old Testament manuscripts is significantly smaller than that of New Testament manuscripts. Nevertheless, when compared to other ancient writings, Old Testament manuscripts are remarkably abundant. The relative scarcity of complete Old Testament scrolls can be attributed to two primary factors. First, the materials used—typically parchment or papyrus—were not durable enough to withstand the passage of two to three millennia without considerable deterioration. Second, it was common scribal practice to destroy the original manuscript once a new, corrected copy was made to replace the aging and damaged text. This practice was motivated by a deep reverence for the sacredness of the Scriptures, ensuring that only pristine copies were preserved and used.

Although the Old Testament is not fully extant in its original Hebrew form, it survives in thousands of fragments and in numerous translations, such as the Greek *Septuagint* and the Syriac *Peshitta*. Unlike the New Testament, which is preserved in a wealth of complete manuscripts, our access to the Old Testament rests on these partial sources. Yet the absence of original-language manuscripts dating close to the time of authorship does not prevent us from reliably reconstructing the original words. One of the strongest forms of bibliographical evidence is the extraordinary care and reverence with

---

[84]See https://www.tandfonline.com/doi/full/10.1080/00206814.2011.639996

which ancient Jewish communities preserved and transmitted their sacred texts.

This deep devotion is reflected in the Talmud[85], a comprehensive compilation of Jewish oral tradition developed from the time of Moses and formally recorded beginning in the second century. The Talmud outlines the meticulous rules that scribes were required to follow when copying the sacred Scriptures. These regulations included strict protocols regarding letter formation, spacing, materials, and even the ceremonial purity of the scribe. Such rigorous standards offer strong assurance that the transmission of the Old Testament was carried out with exceptional precision and dedication across generations. This is an example:

> A synagogue roll must be written on the skins of clean animals, prepared for the particular use of the synagogue by a Jew. These must be fastened together with strings taken from clean animals. Every skin must contain a certain number of columns, equal throughout the entire codex. The length of each column must not extend over less than 48 or more than 60 lines; and the breadth must consist of thirty letters. The whole copy must be first-lined; and if three words be written without a line, it is worthless. The ink should be black, neither red, green, nor any other color, and be prepared according to a definite recipe. An authentic copy must be the exemplar, from which the transcriber ought not in the least deviate. No word or letter, not even a yod, must be written from memory, the scribe not having looked at the codex before him.... Between every consonant the space of a hair or thread must intervene; between every new *parashah*, or section, the breadth of nine consonants; between every book, three lines. The fifth book of Moses must terminate exactly with a line; but the rest need not do so. Besides this, the copyist must sit in full Jewish dress, wash his whole body, not begin to write the name of God with a pen newly dipped in ink, and should a king address him while writing that name he must take no notice of him.

---

[85]There are two known versions of the Talmud: the Jerusalem Talmud (*Talmud Yerushalmi*), composed in the Roman province of Philistia, and the Babylonian Talmud (*Talmud Bavli*), written in the Babylonian region of Mesopotamia. Both versions were developed over several centuries by successive generations of scholars from numerous rabbinical academies established since antiquity.

These were the rules that each of the 304,805 letters of the Pentateuch (the first five books of the Old Testament written by Moses) had to be copied according to.

Among the many Old Testament manuscripts discovered over time, several stand out for their remarkable state of preservation, completeness, and historical significance.

- The *Dead Sea Scrolls*—also known as the *Qumran Scrolls*—represent one of the most significant archaeological finds related to biblical history. Before their discovery, the oldest complete Hebrew Old Testament in our possession was the *Aleppo Codex*, dated to AD 930, while the oldest complete Greek copy was the *Codex Sinaiticus* from around AD 350. Prior to these, only scattered fragments in Hebrew and other languages were available, leaving scholars with little ability to assess the fidelity of these copies to the original texts. The discovery of the *Dead Sea Scrolls* between 1947 and 2017 in the caves near Qumran, along the western shore of the Dead Sea, transformed this landscape. Comprising roughly 40,000 fragments and several dozen complete scrolls—many non-biblical—the collection includes about five hundred reconstructed texts[86]. One of the most significant among them is the complete *Book of Isaiah* (designated 1QIsa), dating to around BC 125. When scholars compared this manuscript to the *Masoretic* text from AD 930, they found only minor discrepancies: of the 166 words in Isaiah 53, only seventeen letters differed—ten were simple copyist errors that did not affect meaning, four were stylistic variations, and three involved the addition of the word "light" in verse 11, a term that also appears in some earlier Greek manuscripts. This extraordinary textual consistency across a millennium strongly supports the reliability and preservation of the Old Testament.
- The *Aleppo Codex*, dated to AD 930, is the earliest known manuscript of the Tanakh (Hebrew Bible) and is considered the most authoritative exemplar of the Masoretic tradition. Written in

---

[86]You can view these scrolls digitally and in extraordinary resolution at: http://dss.collections.imj.org.il/

Hebrew, it reflects the work of the Masoretes—Jewish scribes who succeeded earlier scribes in Tiberias and Jerusalem between the 7th and 10th centuries. Their name derives from the Hebrew word *masoret*, meaning "tradition," underscoring their role in preserving and standardizing the text of the Hebrew Scriptures. Tragically, the codex is now incomplete. During the anti-Jewish riots in Aleppo on December 2, 1947, Arab rioters destroyed numerous synagogues, including the 1,500-year-old Mustaribah Synagogue, where the *Aleppo Codex* had been safeguarded.

- The *Leningrad Codex* belongs to the same Masoretic tradition as the *Aleppo Codex* and is currently the oldest complete manuscript of the Hebrew Bible. It was written in Cairo around AD 1010 and is now housed in the *Russian National Library* in St. Petersburg. Like the *Aleppo Codex*, it features vocalization—the inclusion of vowels—added by the Masoretes to preserve accurate pronunciation and interpretation. Since ancient Hebrew was originally written using only consonants and had fallen out of spoken use by the 4th or 5th century, the addition of vowels became essential. These vowels were inserted with markings to indicate that they were editorial additions and not part of the original consonantal text. The *Leningrad Codex* remains a critical reference for modern editions of the Hebrew Bible and is widely used in biblical scholarship today.

In BC 587, Nebuchadnezzar II, king of Babylon, invaded the Kingdom of Judah, destroyed Solomon's Temple, and carried off the political, religious, and cultural elites into captivity. This period of Babylonian exile lasted for approximately fifty years. In BC 538, the Persian king Cyrus the Great issued a decree that allowed the exiled Jewish families to return to their homeland. However, during their exile, the Jews had become dispersed throughout regions where Greek and Aramaic were the dominant languages. Over time, Greek emerged as the prevailing language of commerce, education, and public life among many Jewish communities in the diaspora, particularly in Egypt.

This linguistic shift created a compelling need to translate the Hebrew Scriptures into Greek. The task of producing this translation was initiated under Ptolemy II Philadelphus, who ruled Egypt in the third century BC. Desiring to include the Jewish Scriptures in the *Library of Alexandria*, one of the ancient world's greatest centers of learning, Ptolemy commissioned his royal librarian, Demetrius of Phalerum, to oversee the translation. According to the *Letter of Aristeas*, Demetrius delegated the task to Aristeas, an Alexandrian Jew, who traveled to Jerusalem to select seventy-two elders—six from each of the twelve tribes of Israel—to perform the translation.

The translation was completed in seventy-two days, after which it was read publicly to the Jewish[87] community in Alexandria, who affirmed its accuracy and sanctity. This Greek translation came to be known as the *Septuagint*, from the Latin *septuaginta* (meaning "seventy"), often abbreviated as LXX, in reference to the seventy or seventy-two translators.

The significance of the *Septuagint* cannot be overstated. It is cited over 250 times in the New Testament, including in the words of Jesus Himself, underscoring its authoritative status among early Christians. Moreover, many of the most important Old Testament manuscripts we possess today—such as the *Codex Alexandrinus*, *Codex Sinaiticus*, and *Codex Vaticanus*—are copies of the *Septuagint*, not the original Hebrew texts.

When assessed using the bibliographical test, the Old Testament demonstrates remarkable reliability. This is evident in the meticulous care with which Hebrew scribes preserved and transmitted the text, the extensive number of surviving manuscripts, and the relatively short time span between the composition of the original texts and our earliest extant copies. The tradition of textual preservation, combined with the widespread use and early citation of the *Septuagint*, affirms the enduring authenticity of the Old Testament Scriptures.

The test of the external evidence of the Old Testament. Archaeology has made invaluable contributions to affirming the historical reliability of the Old Testament by uncovering external evidence that corroborates biblical narratives. Recent excavations near the southern end of the Dead Sea—close to the region historically known

---

[87]Known as Hellenistic Jews.

as the Valley of Sidim—have identified what is believed to be the ancient location of Sodom and Gomorrah. The site aligns precisely with the geographical descriptions found in the Bible. Multiple stratified layers of earth appear to have been violently disrupted and hurled into the air, suggesting that the cities were obliterated by a cataclysmic seismic event. Notably, the region's abundance of bituminous tar supports the biblical account in Genesis 19, which describes fire and brimstone raining down upon the city.

Between 1930 and 1937, archaeologist John Garstang led an extensive excavation of ancient Jericho. His findings are meticulously documented in *The Foundations of Bible History: Joshua, Judges*. Among his most striking discoveries was evidence that the city's walls had collapsed outward—a phenomenon unheard of, as city walls typically fall inward when breached. Garstang remarked:

> As to the main fact, then, there remains no doubt: the walls fell outwards so completely that the attackers would be able to clamber up and over their ruins into the city. Why is it so unusual? Because the walls of cities do not fall outwards, they fall inwards. And yet in Joshua 6,20 we read, 'The wall fell flat. Then the people went up into the city, every man straight before him, and they took the city.' The walls were made to fall outward.

Additional archaeological findings have shed light on the early monarchy of Israel. Saul, the first king, was born in the hill country of Judah, southeast of Hebron, at the stronghold of Gibeah. Excavations in the region reveal that slingshots[88] (or sling weapons) were among the most prominent armaments of the era. This discovery reinforces not only the biblical depiction of David's triumph over Goliath in 1 Samuel 17:49, but also the account in Judges 20:16, which records:

---

[88]The sling is one of humanity's oldest weapons. It consists of two cords or straps attached to a central, flexible pouch that holds a projectile. To use it, the sling is grasped by the ends and swung in a circular motion to build momentum; then, one of the cords is released, launching the projectile at high speed. This allows the projectile to travel great distances with significant impact force. Slings have traditionally been made from a variety of materials, including leather, textile fibers, tendons, and horsehair. The projectiles themselves vary and may include rounded or shaped natural stones, sun-dried or baked clay, and even molded lead.

There were seven hundred chosen men among them who were left-handed. Each of them could sling a stone at a hair and never miss.

Further west, between modern-day Jerusalem and Tel Aviv, lies Tell Gezer—known simply as Gezer during the reign of Solomon[89]. Excavations conducted in 1969 revealed a layer of ash covering most of the city's mound. The site yielded Hebrew, Egyptian, and Canaanite artifacts, indicating the concurrent presence of these cultures, precisely as described in 1 Kings 9:16–17:

> Pharaoh, the king of Egypt, had gone up and captured Gezer. He burned it down and killed the Canaanites who were living there. He gave it as a dowry to his daughter, Solomon's wife. Solomon then rebuilt Gezer.

During the archaeological campaign launched in 2012 at the ancient city of Khirbet Qeiyafa, a remarkable discovery was made in 2015: a ceramic vessel bearing a rare inscription dating back 3,000 years. The inscription mentions Eshba'al Ben Saul, a figure known from biblical tradition as the son of King Saul and a ruler of Israel during the early tenth century BC. This finding provides a significant link between archaeology and the biblical narrative recorded in the Second Book of Samuel, particularly chapters 3 and 4, which recount the complex political and dynastic struggles following Saul's death. The appearance of the name "Eshba'al" on a contemporaneous artifact is especially notable, as it lends historical weight to figures previously known only through scripture.

The city of Shechem—modern-day Nablus, located in the West Bank—boasts a long and layered history. Founded approximately 4,000 years ago in the land of Canaan, Shechem became the first capital of the northern Kingdom of Israel and was associated with the tribe of Manasseh. Its ruins lie about two kilometers east of present-day Nablus and have revealed evidence of the city being destroyed and rebuilt up to twenty-two times, before being firmly reestablished in BC 200.

---

[89]Solomon was the second son born from the union of King David and Bathsheba.

Due to its strategic location in the central hill country of ancient Canaan, Shechem served as a critical commercial hub, trading primarily in grapes, olives, and barley. The Bible frequently references Shechem as a sacred and historical site. In Genesis 12:6, it is recorded that Abram journeyed through the land to Shechem, to the oak of Moreh, at a time when the Canaanites inhabited the region. Later, in Genesis 35:4, the patriarch Jacob is described burying all foreign gods and earrings "under the oak near Shechem," symbolizing a purification ritual and covenantal renewal.

Shechem also holds significance in the New Testament. In his speech recorded in Acts 7:15–16, Stephen, the first Christian martyr, mentions that Jacob and the patriarchs, after dying in Egypt, were brought back and buried in Shechem, in the tomb that Abraham had purchased from the sons of Hamor. This burial narrative further links Shechem to the patriarchal heritage and to the early Christian understanding of divine providence across generations.

Moreover, Joshua 24:32 reinforces Shechem's role in Israel's collective memory:

> The bones of Joseph, that the Israelites had brought up out of Egypt, were buried in Shechem in the parcel of land that Jacob had bought for one hundred pieces of silver from the sons of Hamor, who himself was the father of Shechem. It was an inheritance for the descendants of Joseph.[90]

The Black Obelisk of Shalmaneser III is a monumental artifact from the height of the Assyrian Empire, erected in BC 827 during the reign of King Shalmaneser III[91], who ruled from BC 858 to 824. It was unearthed in 1846 by British archaeologist Austen Henry Layard during his excavation of Nimrud—the ancient Assyrian capital situated on the Tigris River, approximately thirty kilometers southeast of modern-day Mosul, Iraq.

The obelisk is made of black limestone and stands just under two meters tall. Its surfaces are richly carved in high relief, depicting a series

---

[90]His grave can be visited today.

[91]Son and successor of Ashurnasirpal II.

of triumphal scenes that chronicle the king's military campaigns and the tribute he received from vassal states. These tributes include exotic animals such as monkeys, elephants, camels, and rhinoceroses, as well as precious metals, timber, and ivory, showcasing both the wealth of the empire and the extent of its influence.

One of the most remarkable scenes on the obelisk features the earliest known pictorial representation of an Israelite: King Jehu of Israel, who is shown prostrating before the Assyrian monarch. This depiction is especially significant as it confirms a biblical figure in a contemporaneous historical context. The accompanying inscription identifies Jehu as offering tribute to Shalmaneser III, aligning with the political dynamics described in the Hebrew Scriptures.

The biblical narrative of Jehu's rise to power is found in 2 Kings 9:1–3, which describes the divine commissioning of Jehu as king of Israel:

> Elisha the prophet summoned one of the sons of the prophets and said to him, "Gird up your loins and carry this flask of oil to Ramoth-gilead. When you arrive there, search for Jehu, the son of Jehoshaphat, the son of Nimshi. Go to him, and separate him from his companions, bringing him to an inner chamber. Take the flask of oil and pour it on his head, saying, 'Thus says the Lord: I have anointed you as king over Israel.' Then open the door and flee, do not wait around.

Beyond the well-documented discoveries at major excavation sites, an abundance of artifacts spanning various historical periods has been uncovered throughout the ancient Near East. These findings collectively provide a wealth of external corroboration for numerous individuals, cities, and episodes recorded in the Old Testament[92].

Among the most compelling discoveries are those linked to notable biblical figures, such as:

- *The Prophet Balaam*, mentioned in Numbers 22, whose name appears in an inscription known as the Deir 'Alla Inscription, referencing visions and divine communication—an extraordinary parallel to the biblical narrative.

---

[92]See *The Archeology of Ancient Israel*, by Amnon Ben-Tor.

- *Eber*, the patriarch and descendant of Shem, referenced in Genesis 11:15–17, from whom the term "Hebrew" is believed to derive.
- *Goliath of Gath*, the famed Philistine warrior slain by David, as described in 1 Samuel 17:4–23 and 21:9. The city of Gath, his birthplace, has yielded significant Philistine artifacts and was a major urban center during the Iron Age (2 Kings 12:18).
- *Hananiah*, the prophet who opposed Jeremiah, mentioned in Jeremiah 28, a figure entwined in prophetic controversy during Judah's final days.
- *Gemariah*, the son of Shaphan the scribe, who is cited in Jeremiah 36:10 as facilitating the public reading of Jeremiah's scroll in the temple.
- *Jaazaniah*, a military leader during the final years of the Kingdom of Judah (2 Kings 25:23), whose name appears on a seal impression found during excavations.
- *The fortified cities of Lachish and Azekah*, identified in Jeremiah 34:7 as the last strongholds to resist King Nebuchadnezzar II of Babylon before the fall of Jerusalem. Both sites have been extensively excavated, revealing siege ramparts, correspondence tablets (Lachish Letters), and destruction layers consistent with Babylonian conquest.
- *Nineveh*, the capital of the Assyrian Empire and the setting for the prophetic mission of Jonah (Jonah 1:1), whose vast ruins have revealed palatial reliefs, libraries, and inscriptions affirming its grandeur and significance.
- *Belshazzar* (Baltasar), identified in Daniel 5 as the last king of Babylon. While long disputed, his historicity was later confirmed through Babylonian texts identifying him as the co-regent and son of Nabonidus, Babylon's final official monarch.

The proof of the internal evidence of the Old and New Testaments. While it is true that variations exist among biblical manuscripts, such differences are neither surprising nor problematic when we consider the historical context of their transmission. These texts were hand-copied over centuries, and as with any manual reproduction process, copying errors inevitably crept in, often propagating through subsequent copies and gradually diverging from the original. Even in our modern era of printed texts, typographical errors remain common and are typically corrected in later editions.

Recognizing this, scholars have developed the rigorous discipline of textual criticism—a field dedicated to reconstructing the original text by comparing and evaluating the multitude of surviving manuscripts. Through this method, scholars can identify, isolate, and correct errors, assigning greater or lesser textual value to individual manuscripts based on the frequency, type, and severity of deviations they exhibit.

Most manuscript discrepancies fall under the category of unintentional errors. These often include:

- *Confusion of similar-sounding words*, much like "affect" and "effect" in English. In Koine Greek, the language of the New Testament, homophones such as *echoomen* ("have") and *echomen* ("let us have") illustrate the same vulnerability.
- *Omissions*, typically resulting from a scribe inadvertently skipping lines—especially when two lines ended in similar words or phrases.
- *Additions*, often repetitions caused by momentary loss of the scribe's place in the text.

Another common source of confusion came from the practice of marginal notations. Scribes sometimes added explanatory notes or comments in the margins. Over time, as manuscripts were copied and recopied, these notes were occasionally misunderstood as part of the original text and incorporated into the main body, further contributing to textual variation.

Nevertheless, the vast number of extant biblical manuscripts enables scholars to detect these errors with relative ease. As a result, such unintentional variants rarely obscure the meaning of the text and do not significantly compromise its integrity.

More challenging for textual critics is intentional changes—modifications made deliberately by scribes who believed they were correcting what they perceived to be mistakes. In these cases, scholars must attempt to discern the motivations behind the alteration.

A well-known example appears in John 7:39. In early and respected manuscripts such as the *Codex Alexandrinus*, *Codex Vaticanus*, and *Codex Sinaiticus*, the text reads: "for the Spirit was not yet." This phrasing may have troubled some scribes, potentially suggesting that the Holy Spirit did not exist at that time. As a result, some copyists added the word "given", rendering it: "for the Spirit had not yet been given." Others added "Holy", modifying the phrase to "Holy Spirit" for theological clarity.

Such editorial insertions reflect the scribes' efforts to protect the meaning or doctrinal integrity of the text, even if they inadvertently obscured the original wording.

Contrary to popular skepticism, the core message of the Bible has not been corrupted over time. Despite the centuries that separate modern readers from the original manuscripts, the sheer volume of available copies—thousands for the New Testament alone—allows scholars to reconstruct the original with exceptional accuracy.

We can affirm with confidence that the Old Testament, as it exists today, reflects the same text preserved since at least the seventh century BC, and that the New Testament in our possession is virtually identical to what existed in AD 80. No other ancient document possesses greater textual attestation than the Bible, making it the most extensively documented literary work of antiquity.

SECOND THESIS: "IN THE BEGINNING, GOD CREATED HEAVEN AND EARTH"

Apollo 8 marked a pivotal moment in human history as the second staffed mission of NASA's Apollo Space Program and the first crewed spacecraft to leave Earth's orbit, travel to the Moon, orbit it, and return safely. Launched on December 21, 1968, the mission took three days to reach the Moon, where the astronauts spent approximately twenty hours in lunar orbit.

During their historic voyage, the crew delivered a Christmas Eve broadcast that resonated deeply around the world. From the confines of their spacecraft, with the Moon below and Earth suspended in the blackness of space, the astronauts began to read from the opening verses of the Book of Genesis: "In the beginning, God created the heavens and the earth..."

They continued through to the tenth verse, drawing from the most translated, published, and read books in human history. It was a moment of profound significance: a union of scientific achievement and spiritual reflection. As they beheld Earth—an enormous blue sphere speckled with white clouds, green forests, and brown landmasses—the astronauts reminded humanity of its shared origins. In the words of Pope Francis, it was a striking vision of "our common home."

For generations, students were taught that space and time were immutable constants. A meter was always a meter, and a second always a second—whether on Earth or in the farthest corners of the cosmos.

This notion was upended in 1915, when Albert Einstein, during World War I, unveiled his General Theory of Relativity. Einstein demonstrated that space and time—unified as space-time—are not fixed but instead are shaped and distorted by gravity and velocity. Under certain conditions, a meter might no longer measure a meter and a second might no longer take a second. These insights were encapsulated in his groundbreaking field equation, a mathematical representation of how matter and energy warp the fabric of the universe.

Two years later, in 1917, Einstein realized that his equation described a dynamic universe—one capable of expanding or contracting, much like a flexible sheet of rubber. This directly contradicted the prevailing belief of the time, which held that the universe was static, eternal, and unchanging.

In response to this tension, and eager to align with the consensus, Einstein introduced a modification to his theory: the cosmological constant—a term he added to enforce a static model of the universe. With this adjustment, the field equation now described a cosmos with no beginning and no end, which momentarily pleased the scientific community. Einstein himself later admitted this alteration was a compromise, famously referring to it as the "greatest blunder" of his career, especially after subsequent discoveries—such as Edwin Hubble's observations—confirmed that the universe is indeed expanding.

On May 9, 1931, the Catholic priest and astrophysicist Georges Lemaître[93] published a groundbreaking article titled "The Beginning of the World from the Perspective of Quantum Theory" in *Nature*, one of the world's oldest and most respected scientific journals. In this work, Lemaître decisively challenged the long-held belief in a static, eternal universe, a theory supported by leading scientists of his day, including Albert Einstein. Drawing on the implications of general relativity and the emerging principles of quantum mechanics, Lemaître proposed that the universe was not static but expanding and therefore must have had a beginning.

In his hypothesis, he envisioned that if we could reverse the arrow of time, the universe would shrink into a denser and denser state until all matter was compressed into a single point—what he called a "primeval atom." This incredibly dense state would contain all the matter and energy[94] of the present universe. At a particular moment, this "primitive atom" would fragment, initiating the creation of space and time. Lemaître's proposal was revolutionary, laying the conceptual foundation for what would later become known as the Big Bang theory.

Lemaître's ideas were not isolated. A decade earlier, in 1922, the Russian physicist Aleksandr Friedmann[95] developed a mathematical model based on Einstein's own field equations, describing a universe that could expand. Then, in 1929, American astronomer Edwin Hubble presented observational evidence that galaxies are receding from us—demonstrating that the universe is, in fact, expanding. This observation

---

[93]Georges Lemaître (Belgium, 1894-1966) was a Catholic priest of the Jesuit order, as well as a renowned scientist. In remarks to The New York Times, he addressed the apparent duality between faith and science: "I was interested in the truth from the point of view of salvation and from the point of view of scientific certainty. It seemed to me that both paths lead to the truth, and I decided to follow both. Nothing in my professional life, nor in what I have encountered in science and in religion, has ever led me to change my mind."

[94]Einstein's famous formula, $E = mc^2$, expresses the relationship between mass and energy, showing that a small amount of mass can be converted into a large amount of energy.

[95]Aleksandr Aleksandrovich Friedman (Saint Petersburg, June 16, 1888 – Leningrad, September 16, 1925) was a Russian mathematician and meteorologist, best known for his contributions to relativistic cosmology.

provided the empirical confirmation needed to support Friedmann's and Lemaître's theoretical models.

Confronted with this growing body of evidence, Einstein eventually abandoned his "cosmological constant," which he had introduced in 1917 to maintain the idea of a static universe. He would later confess that this adjustment was "the greatest blunder" of his scientific career. The scientific consensus began to shift: the universe had not always existed— it had a beginning.

Long before these scientific breakthroughs, the Judeo-Christian tradition attributed the authorship of Genesis—and the rest of the Pentateuch—to Moses, a towering figure in both faith and history. According to Exodus 6:20, Moses was born to Amram and Jochebed, members of the tribe of Levi, during the time of Pharaoh's decree to kill all newborn Hebrew boys. Placed in a basket and hidden among the reeds of the Nile River[96], Moses was discovered by Princess Termutis, the daughter of Pharaoh, who adopted him and raised him as a prince of Egypt. As the brother of the future pharaoh, Moses received the finest education available in the ancient world.

And yet, the Genesis account he authored contains astonishing statements about the origins of the universe—statements that seem remarkably in harmony with what modern science has only recently come to understand. How could Moses, writing more than three millennia ago, possess knowledge that echoes modern scientific discovery? He described a universe with a definite beginning, an idea confirmed only in the 20th century by cosmologists. He wrote that everything began from nothing, aligning with the theory that space, time, and matter all emerged from a singular point. Moses spoke of the light existing on the first day, before the sun, moon, and stars were created on the fourth day—a detail that resonates with the Big Bang model, where light[97] and energy appeared long before stars formed. He stated that

---

[96]According to what was said in *Jewish Antiquities*, book II, chapter 9, paragraph 5, by Josefo Flavio.

[97]According to the Big Bang theory, the origin of the universe involved a massive explosion that released an immense amount of light and energy. In 1978, the Nobel Prize in Physics was awarded to Arno Penzias and Robert Wilson for their discovery of cosmic microwave background radiation—faint thermal radiation considered a remnant of that

organic matter originated from "soil", in agreement with the law of conservation of energy[98]. He wrote that life began in the oceans, a view now widely held in biology. Most profoundly, Moses reflected the principle of biogenesis—that life only arises from existing life (God)—a foundational concept in biology today.

Space, time, matter, energy, and motion—these five elements govern the entire universe, as evidenced by the general theory of relativity and the evolution of quantum mechanics, disciplines that have earned numerous scientists the Nobel Prize. Remarkably, these foundational elements also appear in the very first verse of Genesis: "In the beginning [time], God created [energy] the heavens [space] and the earth [matter] [...] and the Spirit of God was moving [motion] over the surface of the waters." How could Moses, writing millennia ago, have articulated a vision so aligned with the fundamental structure of the cosmos?

It is important to clarify what it means to "create." A creator brings something into existence from nothing. Human beings transform—crafting furniture from trees and sculptures from stone—but we do not create ex nihilo. The universe, including our planet and the sun, did not emerge from nothing. The Big Bang marked the explosion of essential primordial energy. But where did that energy come from? Only a creator, in the truest sense, could have initiated such an event.

Sacred texts from various world religions also contain accounts of the universe's origins. However, in contrast to the Bible, which speaks of a creation from nothing, many of these texts describe creation from preexisting elements. Let us examine a few examples from religious traditions I consider significant due to the vast number of their adherents.

In Islam, the Qur'an is regarded as the literal Word of God, revealed to the Prophet Muhammad through the archangel Gabriel. These revelations began on December 22 in AD 609, when Muhammad was forty years old, and continued until his death twenty-three years later.

---

primordial event and one of the most significant pieces of evidence supporting the Big Bang model.

[98]Discovered in the mid-19th century, thanks to the work of Julius Mayer, James Prescott Joule, Hermann von Helmholtz, and others.

The Prophet conveyed to his followers the messages he received, which were later compiled into the Qur'an after his death in AD 632. Under the caliphate of Uthman ibn Affan, the Qur'an took its final form: 114 chapters (surahs) composed of verses (ayats). The text is not arranged thematically or chronologically, but generally by the length of the surahs. Consequently, references to Creation are scattered throughout the book. One such passage reads:

> Say, "Do you indeed disbelieve in He who created the earth in two days and attribute to Him equals? That is the Lord of the worlds." And He placed on the earth firmly set mountains over its surface, and He blessed it and determined therein its [creatures'] sustenance in four days without distinction - for [the information] of those who ask. Then He directed Himself to the heaven while it was smoke and said to it and to the earth, "Come [into being], willingly or by compulsion." They said, "We have come willingly." And He completed them as seven heavens within two days and inspired in each heaven its command. And We adorned the nearest heaven with lamps and as protection. That is the determination of the Exalted in Might, the Knowing. (Surah 41.9-12)

Another passage emphasizes the role of water in the origin of life:

> Allah has created every [living] creature from water. And of them are those that move on their bellies, and of them are those that walk on two legs, and of them are those that walk on four. Allah creates what He wills. Indeed, Allah is competent over all things. (Surah 24.45)

From these verses, it is evident that the Qur'an describes the beginning of the universe as originating from "smoke"—a substance already familiar to humankind—and life as arising from a liquid medium.

Hinduism's foundational sacred scriptures are the *Four Vedas*, considered among the oldest religious texts in the world. Each Veda offers unique insights into spiritual practice, cosmology, and the nature of creation:

- *The Rigveda* is primarily a collection of hymns, prayers, and mantras dedicated to various deities and demigods who represent cosmic forces.

- *The Yajurveda* focuses on the performance of religious rituals and sacrificial ceremonies, serving as a manual for conducting rites.
- *The Samaveda*, whose name derives from the Sanskrit word sāman (meaning "song"), contains melodies and chants, often drawing content from the Rigveda but emphasizing their proper musical rendition.
- *The Atharvaveda* includes spells, charms, and ritual practices. Its name, derived from Atharvan (priest), signifies its role in ritual invocation and folk traditions.

These texts were composed orally by priestly poets from various castes between the 14th and 5th centuries BC. Like the Qur'an, the Vedas do not follow a consistent narrative structure. Their hymns are often unordered, repetitive, and occasionally contradictory. In terms of cosmology, the Vedas present multiple accounts of creation, each tied to the deity being worshiped. Below is a summary[99] of the primary creation myths found within Hindu tradition:

- *Brahman – The Supreme Creator.* According to one tradition, Brahman, the divine essence, and creator, emerged from a lotus blossom. He was originally the entire universe and, from himself, created the gods, placing them in their respective realms: Agni (fire) in the earthly realm, Vayu (wind) in the atmosphere, and Surya (sun) in the heavens. He then ascended to Satyaloka, the highest and most exalted sphere, leaving the created universe behind.
- *Vishnu – The Sustainer and Generator.* In another version, Vishnu, alongside his consort Lakshmi, is depicted reclining on a cosmic serpent with a thousand heads. In his form as Narayana, Vishnu endures the destructive fire and flood that preceded cosmic regeneration. From his navel springs a lotus flower, and from this flower Brahma, the creator god, is born. Thus, Vishnu serves as the generative force from which the act of creation proceeds.

---

[99]You can see the books I quoted on the web page https://www.sacred-texts.com/hin/index.htm (Rigveda, book X, hymns 72, 81, 90, 121, 129, 181, 182 and 190).

- *Shiva – The Transformer and Cosmic Architect.* A third account centers on Shiva, who holds a jug made of clay containing the nectar of immortality[100]. This sacred vessel is said to hold the principles of creation—the Vedas themselves. After crafting the jar, Brahman places it in the cosmic waters that cover the earth following a cyclical, regenerative deluge. In his wanderings, Shiva takes the form of a hunter, and by shooting an arrow into the jar, he releases the seeds of creation, thereby initiating a new cosmic cycle. This story reflects Hinduism's cyclical concept of time, in which the universe undergoes repeated destruction and rebirth.

The Rigveda also includes the Purusha Sukta, a profound hymn describing the creation of humanity[101] through the sacrifice of Manu, the primordial being. From this cosmic sacrifice emerged all aspects of existence: from his mouth came the Brahmins (priestly caste), from his arms, the Kshatriyas (warriors), from his legs, the Vaishyas (merchants and farmers) and from his feet, the Shudras (servants and laborers).

Moreover, cosmic elements are said to have emerged from him: the moon from his mind, the sun from his eye, and the wind from his breath.

As these diverse narratives illustrate, Hindu cosmology embraces multiple, often symbolic creation stories, each associated with a different deity. Importantly, in all versions, creation proceeds from preexisting elements—such as a lotus flower, a clay jar, or the body of a primordial being—rather than emerging from nothing.

The *Tripitaka* (also spelled Tipitaka) constitutes the core of the Buddhist canon, compiled during the reign of King Walagambahu in Sri Lanka during the first century BC, approximately five hundred years after the passing of Siddhartha Gautama, the historical Buddha[102]. These

---

[100]In Sanskrit, Amrita—meaning "deathless"—is the name given to the mythical nectar of immortality. The term symbolizes eternal life and divine essence in Hindu mythology. Amrita has been etymologically associated by some with the word Atlantic, which has been interpreted in certain esoteric or symbolic contexts to mean "one who transcends through the inexplicable" or "one who possesses or understands feminine energy."

[101]See https://universohindu.com

[102]Siddhartha Gautama, better known as Gautama Buddha—or simply the Buddha—was a monk, mendicant, philosopher, and sage whose teachings laid the foundation of

scriptures represent the foundational teachings of Theravāda Buddhism and are divided into three primary "baskets" (pitaka):

- *Sutta Pitaka* – A compilation of discourses attributed to the Buddha, covering ethical teachings, meditative practices, and philosophical dialogues.
- *Vinaya Pitaka* – The monastic code outlining rules and ethical conduct for monks and nuns.
- *Abhidhamma Pitaka* – A detailed scholastic analysis of mental processes and phenomena, presenting a systematic interpretation of the Buddha's teachings.

Unlike Abrahamic or Hindu religious texts, the Tripitaka does not contain a creation narrative. Nowhere in the Buddhist scriptures is there a mention of a deity creating the universe or humankind. Instead, the Buddhist worldview emphasizes a cyclical understanding of existence: everything that exists—whether material or immaterial—is subject to birth, life, decay, and death, repeating endlessly across infinite eons. Matter is considered eternal, not created from nothing, but arising and dissolving in accordance with the law of dependent origination (paṭicca samuppāda).

In addressing metaphysical questions, including those concerning the origin of the universe and life, the Buddha advised restraint and focus on liberation, rather than idle speculation. A pivotal teaching in this regard comes from the Acintita Sutta ("The Unconjecturables"), which warns against attempting to fathom that which lies beyond ordinary comprehension:

> These four unconjecturable, oh monks, should not be thought; conjectured of these, one would experience grief and madness. What are these four? (1) The sphere [of knowledge] of the

---

Buddhism. He was born in the ancient Shakya Republic, located in the foothills of the Himalayas, and taught primarily in the northwestern regions of India.

To prevent common misconceptions, it is important to clarify that Gautama Buddha is not considered a god, nor is he the only or the first Buddha. According to Buddhist cosmology, the title "Buddha" refers to one who has attained full enlightenment, a state that any human can achieve. Humans are seen as possessing the greatest potential for enlightenment, though this is not limited to humanity as we know it.

Buddhas, O monks, is an unconjecturable that should not be thought; conjectured about this, one would experience grief and madness. (2) The sphere of meditative absorptions, oh monks! is an unconjecturable that should not be conjectured; conjectured about this, one would experience grief and madness. (3) The result of actions (kamma), oh monks, is an unconjecturable that should not be conjectured; conjectured about this, one would experience grief and madness. (4) *To conjectured about the [origin] of the world, oh monks, is an unconjecturable that should not be conjectured*; conjectured about this, one would experience grief and madness. These four unconjecturable, oh monks, should not be conjectured; conjectured of these, one would experience grief and madness. (Acintita Sutta 392, *Sixth Buddhist Council; emphasis* mine)

This passage underscores a key feature of early Buddhism: its agnostic approach to cosmological origins. The Buddha viewed questions about the beginning or end of the universe as ultimately irrelevant to the cessation of suffering, which is the central concern of the Dharma.

Buddhism offers a strikingly different perspective from many religious traditions. It does not posit a divine creator or an act of creation ex nihilo. Instead, it presents a universe governed by causality, impermanence, and interdependence. The focus is not on how or why the universe came to be, but rather on how beings suffer and how they can attain liberation (nirvāṇa). The Buddha's repeated emphasis on pragmatic insight over metaphysical speculation reinforces this shift from cosmological beginnings to existential ends.

The Bible, though composed more than 3,500 years ago in a language embedded with symbolism and metaphor, continues to astonish scholars and believers alike with its elegance, internal coherence, clarity, and precision. Despite its ancient origins, the biblical narrative of the universe's creation remarkably aligns in key stages with what modern scientific discovery has revealed only in the last century. Let us explore this correspondence in greater depth.

According to modern scientific understanding, the universe began with a monumental event known as the Big Bang—a singularity from which space, time, matter, and energy all emerged simultaneously. This idea of a definitive beginning stands in contrast to earlier scientific assumptions of an eternal, unchanging universe. Yet, the very first verse of the Book of Genesis already states this truth plainly: "In the beginning, God created the heavens and the earth" (Genesis 1:1), declaring a cosmic origin from a moment in time.

Science further explains that this immense explosion released an extraordinary quantity of light and energy, a remnant of which still permeates the cosmos today as cosmic microwave background radiation[103]. The Bible mirrors this phenomenon closely in its next statement: "Then God said, 'Let there be light,' and there was light" (Genesis 1:3). Light, in both accounts, marks one of the very first consequences of creation.

As the universe continued to expand and cool, atoms—primarily hydrogen and helium—formed and began to coalesce into vast, formless clouds. Gravitational forces pulled these particles together, giving rise to the first stars, galaxies, and planetary systems. In the Genesis narrative, we read: "God called the dry ground 'land'" (Genesis 1:10), and shortly after, "Let there be lights in the vault of the sky to separate the day from the night..." (Genesis 1:14–15). Here, the formation of celestial bodies, including stars and planets, appears in a sequence that aligns with scientific understanding.

Modern geology and planetary science explain that Earth's atmosphere began forming about four billion years ago. As the planet cooled after a violent era of volcanic activity, water vapor accumulated and eventually condensed into clouds, leading to the formation of rain and oceans. The Bible's version reads: "There were not yet any plants of the field nor had any herbs sprouted... for the Lord God had not made it rain upon the earth... He made a mist rise from the ground to water the whole surface" (Genesis 2:5–6), a strikingly similar depiction of Earth's early hydrological processes.

As life began to emerge, science tells us that unicellular organisms—notably phytoplankton—were among the earliest living things to appear,

---

[103]The cosmic microwave background radiation was first detected in 1965 by American physicists Arno Penzias and Robert Woodrow Wilson at Bell Laboratories in Crawford Hill, near Holmdel Township, New Jersey. This groundbreaking discovery earned them the Nobel Prize in Physics in 1978.

laying the foundation for the plant kingdom. The Bible also places plant life before animals, stating: "God said, 'Let the land bring forth plants... each according to its own kind.' And it was so." (Genesis 1:11). The sequence is again consistent with the fossil record.

Continuing the narrative, scientific findings show that marine animals preceded terrestrial creatures, first populating the oceans before venturing onto land. The Cambrian Explosion offers strong evidence of this sudden diversification of animal life. Genesis describes the process in much the same order: "God created the great sea creatures and all the other creatures that fill the waters... God said, 'Let the earth bring forth living creatures...'" (Genesis 1:21-24), mirroring the progression from sea to land.

Finally, the human being emerges as the most complex of all creatures. According to science, humans share the same basic organic elements as all other living organisms—derived from the earth itself. The Bible agrees but adds a dimension science cannot quantify: "Then the Lord God formed man out of the dust of the earth, and He breathed into his nostrils the breath of life, and man became a living being." (Genesis 2:7). Here, the biblical account introduces the concept of the soul, the divine breath that imparts consciousness and spiritual identity—a profound mystery that remains beyond the reach of empirical explanation.

Having now examined the cosmological views of the four most widely followed world religions—Islam, Hinduism, Buddhism, and Christianity—a fascinating observation emerges. The Qur'an acknowledges a divine creation but often portrays it as emerging from preexisting elements, such as smoke and water. Hinduism offers multiple narratives, each associated with different deities, often invoking symbolic and material metaphors like the lotus flower or sacred vessels. Buddhism, in contrast, abstains from a creator myth altogether, presenting a vision of cyclical existence and actively discouraging speculation about the origin of the universe.

Among these perspectives, the Biblical account in Genesis stands out not only for its linear chronology but also for its striking alignment with scientific discoveries, described in a manner that is sophisticated, detailed, and uncannily accurate—particularly for a text written in antiquity. Its bold claims about a beginning, the emergence of light, the sequence of cosmic and biological development, and the unique nature of humanity seem too coherent to be attributed solely to ancient imagination.

More than 75% of the global population adheres to one of the four religions discussed. Yet it is the Biblical creation narrative that appears to resonate most closely with the insights of modern science. Could such detailed convergence—written millennia ago by authors without access to telescopes, particle accelerators, or genomic analysis—be chalked up to coincidence or poetic luck?

THIRD THESIS: SCIENTIFIC FACTS IN THE BIBLE

The apostle Paul met Timothy during his second missionary journey to the city of Lystra, located in present-day Turkey. Their relationship quickly deepened into one of companionship and spiritual kinship, with Timothy eventually becoming one of Paul's most trusted allies. Paul would later write two epistles addressed specifically to Timothy, offering both pastoral guidance and personal encouragement. In his second letter, Paul reminds Timothy of the profound importance and role of Scripture:

> But as for you, stand by what you have learned and firmly believed, because you know from whom you have learned it. Gain Wisdom from the Inspired Scriptures. Also remember that from the time you were a child you have known the sacred Scriptures. From these you can acquire the wisdom that will lead you to salvation through faith in Christ Jesus. All Scripture is inspired by God and is useful for teaching, for refutation, for correction, and for training in uprightness, so that the man of God may be proficient and equipped for good work of every kind. (2 Timothy 3:14-17)

In this passage, Paul emphasizes that the Bible's central purpose is spiritual: to impart wisdom, lead to salvation, and provide instruction for righteous living. It is not intended as a manual of scientific explanation. However, that does not mean the Bible is scientifically ignorant or irrelevant. In fact, it contains certain insights and truths that only much later were validated by science, prompting reflection on the depth and foresight of its content.

The biblical authors wrote within their own historical, cultural, and linguistic context. They addressed specific audiences, often limited to their immediate geographical regions and eras, and conveyed divine truths using the tools available to them—their own languages, literary styles, and rhetorical traditions. These authors did not aim to write textbooks on cosmology, biology, or geology; instead, they sought to express theological and moral truths through narrative, poetry, wisdom literature, prophecy, and epistle.

A useful analogy illustrates the nuance of biblical language: imagine yourself sitting peacefully in the passenger seat of a car, enjoying a scenic view under the sun. The calm of the moment might lull you into a gentle daydream. Suddenly, the driver brakes, makes a sharp turn, and accelerates—the abrupt shift demands your full attention. Similarly, figures of speech in Scripture are employed to grab the reader's attention, to create impact, or to draw focus to a key moment or idea.

The Bible makes frequent and masterful use of such rhetorical devices to bring the narrative to life. There are over two hundred figures of speech used across its texts, but among the most prevalent are: Simile[104], Metaphor[105], Allegory[106], Paradox[107], Irony[108],

---

[104]This literary device involves comparing a real concept with an imaginary one that shares a similar quality. Examples of this can be found in Psalm 1:3 and 1 Peter 2:25.

[105]This literary device involves identifying a real concept with an imaginary one based on a shared similarity. Examples of this can be found in Isaiah 40:6, 1 Peter 1:24, Psalm 23:1, Matthew 5:13, and Matthew 26:26.

[106]This literary device consists of a series of metaphors presented in succession, which together evoke a more complex or layered idea. Examples can be found in Galatians 4, Psalm 80, Isaiah 5, and Matthew 12:43–45.

[107]This literary device involves combining two opposing or seemingly contradictory ideas that, when considered together, may reveal a deeper or hidden truth. Examples can be found in Matthew 16:25 and 1 Timothy 5:6.

[108]This literary device involves implying the opposite of what is stated, often for rhetorical or humorous effect. Examples can be found in Job 12:2, 1 Kings 18:27, and Luke 13:33.

Personification[109],   Anthropomorphism[110],   Anthropopathy[111],
Hyperbole[112], Synecdoche[113] and Euphemism[114].

Each of these devices serves a communicative purpose, often aiming to illuminate deeper meaning or evoke emotional resonance. Their use, however, demands that the reader approach Scripture with discernment and care. Failure to distinguish between literal description and figurative language may lead to misinterpretation or confusion. For example, taking anthropomorphic depictions of God too literally can obscure the transcendent nature that Scripture simultaneously affirms.

In the second century, the Greco-Roman mathematician and astronomer Claudius Ptolemy[115] introduced one of antiquity's most influential astronomical treatises: *The Almagest*. In this seminal work, Ptolemy articulated and formalized the geocentric model of the universe—a view that had earlier roots in the philosophies of Plato and Aristotle. According to this model, Earth stood motionless at the center of the cosmos, and all celestial bodies—including the sun, moon, planets, and fixed stars—revolved around it in concentric spheres.

This Earth-centered cosmology was not merely a scientific hypothesis; it soon acquired strong philosophical and theological endorsement, especially from religious authorities who interpreted it as

---

[109]This literary device involves attributing human qualities or actions to animals, objects, or abstract ideas. Examples can be found in Matthew 6:24 and Judges 5:20.

[110]This literary device involves attributing human form or human characteristics to God. Examples can be found in Exodus 33:11, Job 34:21, James 5:4, and Isaiah 30:27.

[111]This literary device involves attributing human emotions or feelings to God. Examples can be found in Genesis 6:6 and Exodus 20:5.

[112]This literary device involves exaggerating or diminishing an aspect or characteristic of something to an extreme degree. Examples can be found in Exodus 8:17, Deuteronomy 1:28, and Judges 20:16.

[113]This literary device involves referring to a part to represent the whole, or the whole to represent a part. Examples can be found in Matthew 6:11 and Proverbs 22:9.

[114]This literary device involves replacing a harsh or unpleasant word or expression with one that has softer or more agreeable connotations. Examples can be found in John 3:16 and Revelation 22:18.

[115]Claudius Ptolemy (c. AD 100 – c. 170), born in Ptolemaida Hermia and later based in Canopus, was a Greek astronomer, astrologer, geographer, mathematician, and possibly a chemist. His works had a profound influence on science and thought throughout antiquity and the Middle Ages.

consistent with Scripture. A frequently cited passage was Psalm 93:1–2, which reads:

> The Lord is King, adorned in splendor; the Lord has clothed and girded himself with strength. He has made the world firm, never to be moved.

The geocentric model, which placed Earth immobile at the center of the universe, was upheld by the Church for many centuries, due to a literal interpretation of scriptural texts such as Psalm 93:1–2. This view remained dominant until 1532, when Nicolaus Copernicus proposed his revolutionary heliocentric theory, asserting that the Earth is in motion, orbiting around a stationary sun.

While the biblical text itself was never in error, the longstanding misinterpretation of metaphorical language was gradually corrected considering new scientific understanding. The psalmist's statement— "He has made the world firm, never to be moved"—was not a commentary on the Earth's physical motion, but rather a metaphorical affirmation of God's sovereign power and stability in creation. It was a poetic expression meant to convey the unshakable nature of God's rule, not a scientific proposition about planetary mechanics.

This example underscores the importance of distinguishing between literal and figurative language when reading the Bible. Many scriptural passages make use of rhetorical devices—such as metaphor, hyperbole, personification, and allegory—to convey spiritual truths and emotional depth. Failing to recognize these literary tools can lead to misunderstandings, especially when theological reflection intersects with scientific discovery.

With this important principle in mind—namely, that not all passages are meant to be interpreted literally—we may now turn our attention to a selection of scientific insights contained within the Bible.

Imagine yourself living in ancient Israel during the time of King David[116], gazing up at the night sky. Without the aid of telescopes or advanced instruments, all you would see are thousands of twinkling

---

[116]He lived between the years BC 1040 and 966.

lights, some brighter than others, scattered across the vast canopy of stars. Could you, with the naked eye alone, confidently declare that each star is fundamentally unique—not merely in brightness or position, but in its very nature? From a purely human perspective of the era, such a claim would be impossible to verify.

And yet, King David, writing over 3,000 years ago, expressed a remarkable insight that resonates with what modern science would only later confirm. In Psalm 147:4, he writes: "He fixes the number of the stars and assigns a name to each."

This verse suggests not only an awareness of the stars' individual identity, but also an astonishing sense of their divine distinctiveness— each one counted, named, and known by the Creator.

Centuries later, the Apostle Paul, in 1 Corinthians 15:41, echoes this same notion: "The sun has a splendor of its own, the moon another splendor, and the stars still another. Indeed, the stars differ among themselves in splendor."

Paul not only recognizes the varying brightness of celestial bodies but goes further in affirming that no two stars are alike in glory, hinting at their intrinsic diversity.

These scriptural observations, which may have seemed poetic or metaphorical at the time, gained empirical validation in the 19th century through the invention of the spectroscope by Joseph von Fraunhofer[117], a German physicist, optician, and astronomer. In 1814, Fraunhofer developed the first spectroscope capable of analyzing the spectral lines of starlight—a breakthrough that revolutionized the field of astronomy.

Through spectroscopy, scientists discovered that each star emits a unique pattern of absorption lines—its spectral "signature"—based on its elemental composition, temperature, and motion. No two stars are identical in their spectra, confirming that each one is indeed distinct, much like a cosmic fingerprint.

---

[117]Joseph von Fraunhofer (Straubing, March 6, 1787 – Munich, June 7, 1826) was a German astronomer, optician, and physicist. He is regarded as one of the founders of spectrometry as a scientific discipline.

To continue with the theme of the stars, would you be willing to assert that their quantity is infinite? Since your eyes can observe a significant number, it would be reasonable to assume you might attempt an estimate. You may suggest there are a thousand, ten thousand, one hundred thousand, or even a million. But would you describe them as infinite?

Just over 2,500 years ago, the prophet Jeremiah made the following statement:

> The word of the Lord came to Jeremiah: Thus says the Lord: [...]
> I will make the descendants of David my servant and the Levites
> who minister before me as countless as the stars in the sky and
> as measureless as the sand on the seashore. (Jeremiah 33:19-22)

Until December 20, 1923, it was widely believed that the Milky Way constituted the entirety of the universe, and that every luminous point in the night sky was simply a stellar object within it. On that day, astronomer Edwin Powell Hubble, observing from the Mount Wilson Observatory in California, made a groundbreaking discovery: one of those points of light—long assumed to be a star—was another galaxy, containing millions of stars.

He observed a second such point and confirmed the same result, then repeated the process with additional points, each time discovering yet another galaxy. Hubble's findings dramatically expanded our understanding of the universe over the following years.

Today, we know that the universe contains not only billions of galaxies, but also an effectively uncountable number of stars—a reality far beyond anything previously imagined.

In Hinduism, our planet is portrayed as a vast serpent that bites its own tail—a clear allusion to the cyclical nature of the universe. In some versions of the tradition, this serpent is suspended in a vacuum, encircling a sea of tranquility composed of milk. Within this cosmic ocean swims a turtle, symbolizing creative power. On the turtle's back stand three elephants, each supporting a world. The lower world is associated with demons and hell, the upper world with gods and prosperity, and the intermediate world, occupied by humans, represents our planet.

The ancient Greeks held a different but equally imaginative cosmology. They believed the Earth was a massive entity supported by columns resting on the shoulders of the Titan Atlas. According to Greek mythology, Atlas had led the Titan rebellion against the Olympian gods—a conflict known as the Titanomachy. As punishment for his defeat, Zeus condemned Atlas to carry the weight of the Earth on his back for eternity.

Maheo, the Great Spirit of the Cheyenne—one of the principal Indigenous peoples of North America—commanded the turtle to bear the world on its shell. This choice symbolized the turtle's strength and longevity[118], traits deeply revered in their tradition.

In contrast to these mythological depictions, the Bible presents a remarkably different view. The Earth does not rest on any animal or physical support but is described as suspended freely in space. This concept is found in the book of Job, which is believed to have been written between the 10th and 8th centuries BC:

> He stretches out the North above the void and suspends the earth on nothingness. He encloses the waters in dense clouds, yet the clouds are not torn asunder under their weight. He veils the face of the full moon, spreading his clouds beneath it. "He has established the horizon on the surface of the waters as the boundary between light and darkness. (Job 26:7-10)

A clear reference to the roundness of the Earth is made by the prophet Isaiah in his description of Creation: "God sits enthroned above the dome of the earth." (Isaiah 40:22)

This imagery suggests a spherical or curved Earth, long before such a concept was widely accepted in science.

Furthermore, the evangelist Luke describes the second coming of Jesus as a sudden and instantaneous global event. His account implicitly

---

[118]The story is found in *Legends of the North American Indians*, by Francisco Caudet Yarza.

acknowledges that day and night occur simultaneously on Earth—a phenomenon only possible on a rotating, spherical planet. Luke writes:

> I tell you, on that night two people will be in one bed. One will be taken and the other will be left. And there will be two women grinding grain together. One will be taken and the other will be left. Two men will be out in the field. One will be taken and the other will be left. (Luke 17:34-36)

This was scientifically verified fifteen centuries later, when renowned navigators such as Christopher Columbus, Vasco da Gama, Pedro Álvarez Cabral, Juan de la Cosa, Bartolomé Díaz, Diego García de Moguer, Ferdinand Magellan, Andrés de Urdaneta, Diego de Almagro, Francisco Pizarro, Francisco de Orellana, and Hernán Cortés circumnavigated and mapped the Earth. Through their voyages, they confirmed that the Earth is suspended in space and spherical in shape.

In the first chapter, I discussed the second law of thermodynamics, also known as the law of entropy, which asserts that matter deteriorates over time. This implies that, given enough time, all matters will decay and vanish. In 1824, the French engineer Nicolas Léonard Sadi Carnot published his work *Reflections on the Motive Power of Fire and on Machines Fitted to Develop That Power*—the first formal articulation of this principle. The theory continued to evolve until the early 20th century, when Albert Einstein introduced his work on Special Relativity, expanding our understanding of energy, matter, and time.

Yet, the Bible had already conveyed the principle of universal decay. The prophet Isaiah and King David both warned of the Earth's inevitable erosion, expressing a truth that would not be fully grasped for over two millennia.

> Raise your eyes to the heavens and gaze down on the earth below. For the heavens will vanish like smoke, and the earth will wear out like a garment as its inhabitants die like flies. But my salvation will be everlasting and my justice will never cease. (Isaiah 51:6)

> Long ago you laid the foundations of the earth, and the heavens are the work of your hands. They will pass away but you endure;

they will all wear out like a garment. You will change them like clothing, and they will perish. (Psalms 102:26-27)

In the 1930s, James Jeans, an English physicist, mathematician, and astronomer, proposed the Steady State Hypothesis. This theory suggested that matter was being continuously created to account for certain cosmological phenomena that could not be explained by existing models at the time. However, this idea directly contradicted the first law of thermodynamics, which states that matter and energy can neither be created nor destroyed.

In contrast, the Bible had long affirmed a completed creation. As written in Genesis 2:1: "Heaven and earth, and everything that is in them, were finished."

FOURTH THESIS: THE PROPHECIES FULFILLED IN JESUS

In the year 2000, a friend[119] of mine made a surprising promise: he would predict a future event that I would later be able to verify. Naturally skeptical, I urged him to write it down, assuming it was just a game. He handed me a sealed envelope with the words "Open on January 1, 2020" written on the front.

When the day finally arrived, I opened the envelope with curiosity, having kept that date in mind for two decades. The message inside read: "On this date, January 1, 2020, a child is to be born at Mount Sinai Hospital in New York City."

At first, I was unimpressed. Anyone could have made such a generic statement, and the odds were high that a child would indeed be born there on that date. Still, I called the hospital to verify—and, as expected, a boy had been born that day. But did this make my friend a "prophet"? Of course not. As I said, anyone could have predicted that.

Now imagine the letter had gone further: "A child is due at Mount Sinai Hospital in New York City on this date, January 1, 2020, and the mother's name is Rosalba." If I had called the hospital and learned that someone named Rosalba had indeed given birth that day, I would have

---

[119]This is a fictional character that I use to explain my point.

been impressed—but not convinced of prophecy. After all, it is plausible that someone with that name gave birth there.

But what if the letter had said this: "A child is to be born at Mount Sinai Hospital in New York City on January 1, 2020. His father's name is Carlos Martínez, and his mother's name is Rosalba Pérez. He is Venezuelan and she is Ecuadorian. It is their first child. Carlos is thirty years old, and Rosalba is twenty-four. The child will be named Felipe."

And what if the hospital confirmed that Felipe, the couple's first child, had indeed been born that day to Rosalba Pérez, a 24-year-old Ecuadorian woman, and Carlos Martínez, a 30-year-old Venezuelan man?

At that point, two explanations would remain: My friend truly had the ability to predict the future, or he randomly guessed a highly specific sequence of names, nationalities, ages, relationships, and events—and somehow got it all exactly right.

But how likely is it that he simply fabricated all that detailed information… and happened to be right on every count?

I provide a brief overview of the fascinating field of probability in Appendix B but understanding just how difficult it is to accurately guess all the details from my previous example does not require advanced mathematics. Most people are familiar with lottery, which serve as a relatable illustration.

Imagine a raffle with only nine tickets. Winning in that case would seem quite easy. Now, increase the number to ninety-nine tickets. The chances of winning suddenly diminish. And if there were 999,000 tickets, it would become extraordinarily difficult to win—almost impossible, in fact. The larger the pool, the smaller the probability of selecting the correct outcome by chance.

What my friend did when he made his prediction was something much more complex than simply picking a winning number. He selected one city out of all the cities in the world. He specified a single date from an enormous range of possible dates. He named two individuals, choosing a man and a woman from countless possible names. He gave their ages, their nationality, and even the name that would be given to their child. Each of these choices, taken alone, would already be unlikely to guess correctly. Taken together, the odds become astronomically small—so small that the idea of it being a coincidence becomes almost impossible to accept.

If every detail in that prediction turned out to be accurate, then there would be only one reasonable conclusion: my friend was not guessing. He had somehow foreseen an event twenty years into the future and committed it to writing. The only explanation left is that he possessed an extraordinary ability—something that could rightly be called *prophetic*.

The case of Jesus of Nazareth followed a strikingly similar pattern. Over the span of hundreds of years, numerous prophets provided detailed information that all pointed toward a single person: the Messiah. These individuals had never met each other. Many lived on different continents, spoke different languages, and belonged to different historical periods, yet they consistently conveyed prophecies that aligned with the life of one man. They foretold details about his birthplace, the timing of his arrival, the identity of his parents, key events surrounding his life, his companions and enemies, his miracles, his actions, the way he would die, the betrayal by Judas, the desertion of his apostles, and even his resurrection, along with many other aspects of his existence.

Can all of this be dismissed as coincidence? Was it simply good fortune? Or does it point to a far deeper truth—one that reveals who truly authored the Bible?

Many people associate prophets primarily with their role in predicting future events, but it is essential to understand the full scope of their mission. While foretelling was certainly part of their responsibility, it was by no means their primary role, nor their most important one. The core of their calling was spiritual: to guide the people in placing their trust fully in God and to urge them to remain faithful to His covenant. Whether delivering divine instruction or issuing warnings to Israel about the consequences of turning to other gods, the prophets were deeply involved in every stage of Israel's spiritual journey.

Their messages were always rooted in the historical realities of their time. They encouraged the people to uphold their covenant with the Most High, even during hardship, and constantly warned against the seductions of polytheism—which was widespread among the neighboring nations. Pagan rituals, idolatry, and syncretism posed constant threats to the purity of Israel's worship, and the prophets stood as relentless voices calling the people back to true devotion.

Because of their uncompromising stance against anything contrary to God's plan—whether it was social injustice, immorality, corruption, or idolatry—they often found themselves in conflict with kings, priests, and religious authorities, many of whom had grown lax or deliberately ignored the laws that had been given through Abraham's covenant.

Despite their sacred role, many prophets had everyday professions. Jeremiah and Ezekiel were priests. Moses and Amos were shepherds. Deborah served as a judge, Ezra was a teacher, Daniel a royal counselor, Nehemiah a cupbearer to the king, and Job a herder. Yet all of them shared a common duty: to fulfill their earthly responsibilities while proclaiming God's will and exposing any deviation from His divine purpose. They were not simply predictors of the future—they were guardians of truth, messengers of righteousness, and witnesses to the holiness of God in a world filled with compromise.

Being a prophet in Old Testament times was an extremely dangerous calling. The Jewish people understood well the serious consequences for anyone who falsely claimed to speak on behalf of God. According to God's own warning, a death sentence awaited those who were found to be false prophets. Those who prophesied were warned:

> The Lord, your God, will raise up from among your countrymen a prophet who will do what I have done for you, and you will listen to him. This is just as you asked the Lord, your God, at Horeb on the day of the assembly when you said, "Let me not hear the voice of the Lord, my God, anymore, nor look upon this great fire, lest I die." The Lord said to me, "They have spoken well. I will raise up a prophet from among their countrymen who will be like you. I will place my words in his mouth, and he will tell them all that I command him. I myself will call to account whoever does not heed my words that he will proclaim in my name. But if a prophet presumes to proclaim something in my name that I have not said to him, or he speaks in the names of other gods, that prophet is to be put to death." You might say to yourself, "How can we know that the Lord did not speak the message?" If what the prophet proclaims in the name of the Lord

is not true and it does not happen, then the message was not proclaimed by the Lord. The prophet has spoken presumptuously; you should not fear him. (Deuteronomy 18:15-22)

Earlier in this chapter, I provided sufficient evidence that the current Bible can be compared with ancient papyri—or fragments of them—dating as far back as the eighth century BC. This confirms that the Old Testament, as we have it today, is the same text that existed at least eight hundred years before the birth of Jesus.

Why is this important? Because in the following sections, I will be quoting several Old Testament prophecies and explaining how they were fulfilled. I want to eliminate any possibility that someone might claim the prophetic texts were written after the events took place, to fabricate prophecy and falsely prove that Jesus was the Messiah.

That accusation is simply not true. The prophetic writings genuinely predate the birth of Christ by centuries. This is a well-documented fact, and you can verify it through the historical and archaeological sources I previously referenced—some of which are available through reputable academic and historical websites online.

The Bible does not claim that Jesus' apostles were scholarly experts in all the Scriptures—what we now refer to as the Old Testament. However, they were certainly familiar with its first five books, known as the Pentateuch or the Torah, which they referred to as "the Law." Among the twelve apostles, only John and Matthew wrote Gospels, while John, James, and Peter wrote epistles. In all these writings, they emphasized the continued relevance and importance of the Law.

On the day of the resurrection, two disciples encountered the risen Jesus on the road. After their meeting, they reflected on the experience and said to each other: "Wasn't it true that our hearts were on fire when He talked to us on the road and taught us from the Bible?" (Luke 24:32)

What was it that Jesus said that moved them so deeply? What truth did He share that ignited such passion in their hearts?

Jesus must have revealed to them many—perhaps all—of the prophecies that had been written centuries before his birth, all pointing to the Messiah. He showed how these prophecies were fulfilled in his life, death, and resurrection. This profound understanding became so essential to the disciples that, when the evangelists later wrote their Gospels, they saw it as their sacred responsibility to pass on the knowledge Jesus had shared with them personally.

Through this, anyone—even without prior knowledge of Scripture— could examine the prophecies, compare them with the life of Jesus, and conclude that He truly was the Messiah foretold by the prophets.

I am now going to quote several biblical passages that support the fulfillment of Old Testament prophecies. In the vast majority of these, you will notice recurring phrases such as: "this happened so that the Scripture that says [...] might be fulfilled," or "but this happened to fulfill the word that is written in the Law [...]", or "then what was said by the prophet [...] was fulfilled," and "all this has happened so that the Scriptures of the prophets may be fulfilled," or simply "because it is written [...]."

The evangelists deliberately used such language to make it clear to readers that the events they were recording were not random or coincidental. Rather, these events represented the fulfillment of specific prophecies spoken by the prophets long before. Their intention was to help us recognize the deeper meaning behind these occurrences and to show that Jesus was indeed the Messiah, foretold in the Scriptures.

Prophecy one: The Messiah would be the son of God. With this prophecy, Judaism would be the only religion that would proclaim God made man.

Prophecy	Fulfilment
I will proclaim the decree of the Lord: He said to me, "You are my son; this day I have begotten you." Simply make the request of me, and I will give you the nations as your inheritance, and the ends of the earth as your possession. You will rule them with an iron	After Jesus had been baptized, as He came up from the water, suddenly the heavens were opened and He beheld the Spirit of God descending like a dove and alighting on him. And a voice came from heaven, saying, "This is my beloved Son, in whom

scepter; you will shatter them like a potter's vessel. (Psalms 2:7-9).	I am well pleased." (Matthew 3:16-17).
When your days have been fulfilled and you go to be with your fathers, I will raise up your seed after you, one of your sons, and I will establish his kingdom. He will build a house for me, and I will establish his throne forever. I will be his father, and He will be my son. I will not withdraw my mercy from him, as I took it away from the one who preceded you. I will have him stand firm in my house forever, and his throne will be established forever. (1 Chronicles 17:11-14).	

Prophecy Two: He would be born of a woman, which implies that it would not simply appear "out there" without any knowledge of its origin. He would be as human as any of us in the flesh. Mary and her descendant, Jesus, would be the woman of the prophecy.

Prophecy	Fulfilment
The Lord God said to the serpent, "Because you have done this, you will be the most cursed of all the animals and of all the wild beasts. On your belly you shall crawl and you shall eat dust for all the days of your life. I will establish hostility between you and the woman, between your line and her line. Her offspring will crush your head and you will	The birth of Jesus Christ occurred in this way. When his mother Mary was engaged to Joseph, but before they came to live together, she was found to be with child through the Holy Spirit. (Matthew 1:18).

bruise his heel." (Genesis 3:14-15).	

Prophecy Three: He would be born to a virgin, meaning that her pregnancy would not be the result of a relationship with a male, as she would conceive without losing her virginity. I dedicated an entire chapter to this enigma in my debut book, *What You Wanted to Know About the Catholic Church but Were Afraid to Ask*.

Prophecy	Fulfilment
Therefore, you will be given this sign by the Lord himself: The virgin will be with child, and she will give birth to a son, and she will name him Immanuel. (Isaiah 7:14).	The birth of Jesus Christ occurred in this way. When his mother Mary was engaged to Joseph, but before they came to live together, she was found to be with child through the Holy Spirit. (Matthew 1:18).

Prophecy Four: He would be a descendant of Abraham.

Prophecy	Fulfilment
The Lord said to Abram, "Leave your country, your people, and the house of your father, and go to the land to which I will lead you. "I will make of you a great people and I will bless you. I will make your name great and it will become a blessing. I will bless those who bless you and curse those who curse you. And through you all the nations on the earth shall be blessed." (Genesis 12:1-3).	The account of the genealogy of Jesus Christ, the son of David, the son of Abraham. (Matthew 1:1).

Prophecy five: Two of Abraham's offspring were particularly noteworthy: Isaac and Ishmael. Esau and Jacob were the twin offspring of the latter. Jacob was the father of twelve offspring, from whom the twelve tribes of Israel were descended (Genesis 32:28—God changed Jacob's name to Israel). The Messiah would be a descendant of Judah, the fourth of the twelve sons.

Prophecy	Fulfilment
The scepter shall not depart from Judah nor the mace from between his feet, until it comes to whom it belongs, and the obedience of the peoples is his. (Genesis 49:10).	The account of the genealogy of Jesus Christ, the son of David, the son of Abraham. Abraham was the father of Isaac, Isaac the father of Jacob, Jacob the father of Judah and his brothers. Judah was the father of Perez and Zerah, with Tamar being their mother. Perez was the father of Hezron, Hezron the father of Ram. (Matthew 1:1-3).

Prophecy Six: He would be a descendant of Jesse, the father of King David.

Prophecy	Fulfilment
A shoot will spring forth from the stump of Jesse, and a branch will grow from his roots. The Spirit of the Lord will rest upon him: Spirit of wisdom and understanding, a Spirit of counsel and power, a Spirit of knowledge and fear of the Lord, (Isaiah 11:1-2).	The account of the genealogy of Jesus Christ, the son of David, the son of Abraham. [...] Obed was the father of Jesse, and Jesse was the father of King David. (Matthew 1:1-6).

Prophecy Seven: He would be a descendant of King David. The Messiah was prophesied to be the youngest of Jesse's eight sons and a descendant of David.

Prophecy	Fulfilment
Behold, the days are coming, says the Lord, when I will raise up a righteous branch from the line of David. He will reign as king and rule wisely and ensure justice and righteousness in the land. (Jeremiah 23:5).	The account of the genealogy of Jesus Christ, the son of David, the son of Abraham. (Matthew 1:1).

Prophecy eight: He would be born in the city of Bethlehem.

Prophecy	Fulfilment
But from you, O Bethlehem Ephrathah, among the tiniest of the clans of Judah, from you will come forth for me one who is to be a ruler in Israel, one whose origins are from the distant past, from ancient times. (Micah 5:1).	After Jesus had been born in Bethlehem of Judea during the reign of King Herod. (Matthew 2:1).

Prophecy Nine: Kings would travel from distant lands to present gifts to the Messiah.

Prophecy	Fulfilment
The kings of Tarshish and the Islands will offer him tribute; the kings of Sheba and Seba will present him with gifts. (Psalm 72:10). Droves of camels will cover your land, the young camels from	After Jesus had been born in Bethlehem of Judea during the reign of King Herod, wise men traveled from the east and arrived in Jerusalem, [...] And behold, the star that they had seen at its rising proceeded

Midian and Ephah; all from Sheba will come, laden with gold and frankincense, while the people proclaim the praises of the Lord. (Isaiah 60:6).	ahead of them until it stopped over the place where the child was. [...] Then they opened their treasure chests and offered him gifts of gold, frankincense, and myrrh. (Matthew 2:1-11).

Prophecy ten: When monarch Herod heard the rumors regarding the birth of the Messiah, who would become the monarch of Israel, he would commit the slaughter of children under the age of two.

Prophecy	Fulfilment
Thus says the Lord: A voice is heard in Ramah marked by lamentation and bitter weeping. Rachel is mourning for her children, and she refuses to be consoled because they are no more. (Jeremiah 31:15).	When Herod realized that the wise men had deceived him, he flew into a rage and issued an order to kill all the boys in Bethlehem and the surrounding area who were two years old or less, in accordance with the information that he had obtained from the wise men. (Matthew 2:16).

Prophecy eleven: He would be called the Lord.

Prophecy	Fulfilment
The Lord says to my Lord: "Sit at my right hand until I have made your enemies a footstool for you." (Psalm 110:1).	But the angel said to them, "Do not be afraid, for I bring you good news of great joy for all the people. 11 For this day in the city of David there has been born to you a Savior who is Christ, the Lord. (Luke 2:10).

Prophecy twelve: He would be referred to as Emmanuel, which translates to "God with us." In other words, He would be a human being with flesh and blood.

Prophecy	Fulfilment
Therefore, you will be given this sign by the Lord himself: The virgin will be with child, and she will give birth to a son, and she will name him Immanuel. (Isaiah 7:14).	Fear seized all who were present, and they glorified God, saying, "A great prophet has risen among us," and "God has visited his people." (Luke 7:16).

Prophecy Thirteen: He would be recognized as a prophet.

Prophecy	Fulfilment
The Lord said to me, "They have spoken well. I will raise up a prophet from among their countrymen who will be like you. I will place my words in his mouth, and He will tell them all that I command him. (Deuteronomy 18:17-18).	And when He entered Jerusalem, the whole city was filled with excitement. "Who is this?" the people asked, and the crowds replied, "This is the prophet Jesus from Nazareth in Galilee." (Matthew 21:10-11).

Prophecy Fourteen: He would be recognized as a high priest.

Prophecy	Fulfilment
The Lord has sworn, and He will not retract his oath: "You are a priest forever according to the order of Melchizedek." (Psalm 110:4).	Therefore, holy brethren, who share in a heavenly calling, concentrate your thoughts on Jesus, the apostle and the high priest of our profession of faith. (Hebrews 3:1).

Prophecy fifteen: He would be recognized as king.

Prophecy	Fulfilment
I myself have anointed my king on Zion, my holy mountain. (Psalm 2:6).	Above his head was inscribed the charge against him: "This is Jesus, the King of the Jews." (Matthew 27:37).

Prophecy sixteen: An emissary would be responsible for making the announcement of the Messiah's arrival. This individual is named John the Baptist.

Prophecy	Fulfilment
Behold, I am sending my messenger to prepare the way before me. And suddenly the Lord whom you seek will come to the temple, as well as the messenger of the covenant in whom you delight. Indeed, He is coming, says the Lord of hosts. (Malachi 3:1). A voice cries out: In the wilderness prepare the way of the Lord; make a straight path in the desert for our God. (Isaiah 40:3).	In those days, John the Baptist appeared in the desert of Judea, preaching: "Repent, for the kingdom of heaven is close at hand." (Matthew 3:2).

Prophecy Seventeen: His ministry would begin in the Galilee region.

Prophecy	Fulfilment
But there will be no gloom for those who were in anguish. In the former time he brought into contempt the land of Zebulun and the land of Naphtali, but in the	When Jesus learned that John had been arrested, He withdrew to Galilee. Departing from Nazareth, He settled in Capernaum by the sea, in the

latter time He will make glorious the way of the sea, the land beyond the Jordan, Galilee of the nations. (Isaiah 9:1).	region of Zebulun and Naphtali, [...] From that day forward Jesus began to proclaim the message: "Repent, for the kingdom of heaven is close at hand." (Matthew 4:12-17).

Prophecy Eighteen: He would perform numerous miracles and cure an infinite number of illnesses.

Prophecy	Fulfilment
Then the eyes of the blind will be opened and the ears of the deaf will no longer be sealed. Then the lame will leap like a stag and the tongue of the dumb will shout joyfully. (Isaiah 35:5-6).	Jesus traveled through all the towns and villages, teaching in their synagogues, proclaiming the good news of the kingdom, and curing every kind of illness and disease. (Matthew 9:35).

Prophecy nineteen: His preaching would be in the form of parables.

Prophecy	Fulfilment
I will open my mouth in parables and expound the mysteries of the past. These things we have heard and know, for our ancestors have related them to us. (Psalm 78:2-3).	Jesus told the crowds all these things in parables. Indeed, He never spoke to them except in parables. (Matthew 13:34).

Prophecy Twenty: He would enter Jerusalem mounted on a donkey and be proclaimed king.

Prophecy	Fulfilment
Rejoice with all your heart, O daughter Zion. Shout for joy, O daughter Jerusalem. See, your king is coming to you, triumphant and victorious, humble and riding on a donkey, on a colt, the foal of a donkey. (Zechariah 9:9).	Then they brought the colt to Jesus, and after spreading their cloaks over the colt, they helped Jesus to mount it. As he rode along, people kept spreading their cloaks on the road. And when he approached the downward path of the Mount of Olives, the entire multitude of his disciples began to praise God joyfully with a loud voice for all the mighty works they had seen him perform, (Luke 19:35-37).

Prophecy twenty-one: He would not remain in a state of death; rather, He would resurrect. This enigma was the subject of an entire chapter in my debut book, *What You Wanted to Know About the Catholic Church but Were Afraid to Ask.* The entire third chapter of this work will later coalesce around this critical issue, which is a cornerstone of our religion.

Prophecy	Fulfilment
For you will not abandon me to the netherworld or allow your Holy One to suffer corruption. (Psalm 16:10).	But the angel said to the women, "Do not be afraid! I know that you are looking for Jesus who was crucified. He is not here, for He has been raised, as He promised He would be. Come and see the place where He lay. (Matthew 28:5-6).

Prophecy twenty-two: One of his closest friends, the apostle Judas, would be the one to betray him.

Prophecy	Fulfilment
Even my friend whom I trusted, the one who dined at my table, has risen up against me. (Psalm 41:10). And if anyone asks him, "What are these wounds on your chest?" he will reply, "I received them in the house of my friends." (Zechariah 13:6).	While He was still speaking, Judas, one of the Twelve, arrived. With him there was a large crowd of men, armed with swords and clubs, who had been sent by the chief priests and the elders of the people. Now his betrayer had agreed with them on a signal, saying, "The one I shall kiss is the man. Arrest him." Proceeding directly to Jesus, he said, "Greetings, Rabbi!" and kissed him. (Matthew 26:47-49).

Prophecy Twenty-Three: The traitor would receive thirty pieces of silver in return.

Prophecy	Fulfilment
Therefore, it was annulled on that day, and the dealers who were watching me realized that this was the word of the Lord. I said to them, "If it seems right to you, give me my wages; if not, then forget about it." Then they weighed out my wages, thirty pieces of silver. (Zechariah 11:11-12).	[...] and asked, "What are you willing to give me if I hand him over to you?" They paid him thirty pieces of silver, and from that moment he began to look for an opportunity to betray him. (Matthew 26, 15-16).

Prophecy Twenty-Four: That money would be thrown into the temple.

Prophecy	Fulfilment
However, the Lord said to me, "Throw it into the treasury—the princely sum at which they valued my efforts." Therefore, I took the thirty pieces of silver and threw them into the treasury of the house of the Lord. (Zechariah 11:13).	Flinging the silver pieces into the temple, he departed. Then he went off and hanged himself. (Matthew 27:5).

Prophecy Twenty-Five: During his simulated trial, sentence, and execution, his disciples would abandon him.

Prophecy	Fulfilment
Awake, O sword, against my shepherd, against the man who is my associate, says the Lord of hosts. Strike the shepherd, so that the sheep may be scattered, and I will turn my hand against their young (Zechariah 13:7). On that day, every prophet will be ashamed to relate his own prophetic vision, and he will not wear a hairy mantle in order to deceive. Rather, he will say, "I am no prophet. I am a tiller of soil, for the land has been my possession since my youth." And if anyone asks him, "What are these wounds on your chest?" he will reply, "I received them in the house of my friends." (Zechariah 13:4-6).	Then everyone deserted him and fled. (Mark 14:50).

Prophecy Twenty-Six: At the supposed trial, He would be accused by false witnesses.

Prophecy	Fulfilment
False witnesses step forward and question me about things I do not know. (Psalm 35:11).	The chief priests and the whole Sanhedrin tried to elicit some false testimony against Jesus so they could put him to death, (Matthew 26:59).

Prophecy Twenty-Seven: He would not defend himself during the simulated trial; rather, He would remain silent.

Prophecy	Fulfilment
Although harshly treated and afflicted, He did not open his mouth. Like a lamb led to the slaughter and like a sheep that keeps silent before its shearers, He did not open his mouth. (Isaiah 53:7).	But He did not offer a single word in response, much to the governor's amazement (Matthew 27:14).

Prophecy Twenty-Eight: He would endure severe torture, have his face spat upon, and be pummeled to a pulp.

Prophecy	Fulfilment
I offered my back to those who struck me, my cheeks to those who plucked my beard. I did not shield my face from insults and spitting. (Isaiah 50:6).	Then they spat in his face and struck him with their fists. Some taunted him as they beat him, (Matthew 26:67).
But He was pierced for our offenses and crushed for our iniquity; the punishment that made us whole fell upon him, and by his	They also spat upon him and, taking the reed, used it to strike him on the head. (Matthew 27:30).

bruises we have been healed. (Isaiah 53:5). My knees are weak from fasting; my flesh is wasting away. (Psalm 109:24).	He then released Barabbas to them, and after Jesus had been scourged, he handed him over to be crucified. (Matthew 27:26).

Prophecy twenty-nine: He would be mocked by many during his passion.

Prophecy	Fulfilment
But I am a worm and not human, scorned by people and despised by my kinsmen. All who see me jeer at me; they sneer in mockery and toss their heads. (Psalm 22:7-8).	and after twisting some thorns into a crown, they placed it on his head and put a reed in his right hand. Then, bending the knee before him, they mocked him, saying, "Hail, King of the Jews!" (Matthew 27:29). Those people who passed by jeered at him, shaking their heads and saying, "You who claimed you could destroy the temple and rebuild it within three days, save yourself! If you truly are the Son of God, come down from the cross!" (Matthew 27:39-40).

Prophecy thirty: As a result of his crucifixion, his hands and feet would be punctured.

Prophecy	Fulfilment
A pack of dogs surrounds me; a band of evildoers is closing in on	When the other disciples told him, "We have seen the

me. They have pierced my hands and my feet. (Psalm 22:17).	Lord," he replied, "Unless I see the mark of the nails on his hands and put my finger into the place where the nails pierced and insert my hand into his side, I will not believe." (John 20, 25).

Prophecy thirty-one: Accompanied by criminals, He would be crucified.

Prophecy	Fulfilment
Therefore, I will allot him a portion among the great, and He will divide the spoils with the mighty, because He exposed himself to death and was counted among the transgressors, even though He bore the sins of many and interceded for the transgressors. (Isaiah 53:12).	Two thieves were crucified with him, one on his right and the other on his left. (Matthew 27:38).

Prophecy thirty-two: He would intercede for his transgressors during his passion.

Prophecy	Fulfilment
Therefore, I will allot him a portion among the great, and He will divide the spoils with the mighty, because He exposed himself to death and was counted among the transgressors, even though He bore the sins of many and interceded for the transgressors. (Isaiah 53:12).	hen Jesus said, "Father, forgive them, for they do not know what they are doing." (Luke 23:34).

Prophecy thirty-three: He would be rejected by his own people.

Prophecy	Fulfilment
He was despised and shunned by others, a man of sorrows who was no stranger to suffering. We loathed him and regarded him as of no account, as one from whom men avert their gaze. (Isaiah 53:3).	For not even his brethren believed (John 7:5).

Prophecy thirty-four: He would be hated for no reason.

Prophecy	Fulfilment
More numerous than the hairs of my head are those who hate me for no reason. Many are those who seek to destroy me, and they are treacherous. How can I restore what I have not stolen? (Psalm 69:5).	If the world hates you, be aware that it hated me before it hated you. [...] Whoever hates me hates my Father also. If I had not done works among them that no one else had ever done, they would not be guilty of sin. But now they have seen and hated both me and my Father. (John 15:18-24).

Prophecy thirty-five: His acquaintances and associates would withdraw from him and establish a distance.

Prophecy	Fulfilment
My friends and companions stay away from my affliction, and my neighbors keep their distance. (Psalm 38:12).	However, all his acquaintances, including the women who had followed him from Galilee, stood at a distance and watched all these events. (Luke 23:49).

Prophecy thirty-six: They would take away her dress and draw lots for it.

Prophecy	Fulfilment
They divide my garments among them, and for my clothing they cast lots (Psalm 22:18).	When the soldiers had crucified Jesus, they took his clothes and divided them into four shares, one share for each soldier. They also took his tunic, which was woven seamless, top to bottom. They said to one another, "Instead of tearing it, let us cast lots for it to see who is to get it." In this way, the Scripture was fulfilled that says, "They divided my garments among them, and for my clothing they cast lots." (John 19:23-24).

Prophecy thirty-seven: He would experience intense dehydration during his martyrdom, and in lieu of water, they would administer gall with vinegar.

Prophecy	Fulfilment
They put gall in my food, and in my thirst, they gave me vinegar to drink. (Psalm 69:21).	After this, aware that everything had now been completed, and in order that the Scripture might be fulfilled, Jesus said, "I thirst." A jar filled with sour wine was standing nearby, so they soaked a sponge in the wine on a branch of hyssop and held it up to his lips. (John 19:28-29).

Prophecy thirty-eight: His bones would not be fractured after his death, as was the customary practice to guarantee death for victims who had endured the lengthy crucifixion.

Prophecy	Fulfilment
He watches with care over all his bones; not a single one will be broken. (Psalm 34:20).	However, when they came to Jesus and saw that He was already dead, they did not break his legs, (John 19:33).

Prophecy thirty-nine: They would pierce his side.

Prophecy	Fulfilment
Further, I will pour out a spirit of grace and supplication on the house of David and on the inhabitants of Jerusalem so that they will look on me, the one whom they have pierced, and mourn for him as one mourns for an only son, and they will grieve over him as one grieves over a firstborn. (Zechariah 12:10).	But one of the soldiers thrust a lance into his side, and immediately a flow of blood and water came forth. (John 19:34).

Prophecy forty: A great darkness would cover the earth during the martyrdom of Jesus.

Prophecy	Fulfilment
On that day, says the Lord God, I will make the sun go down at noon and darken the earth in broad daylight. (Amos 8:9).	Beginning at midday, there was darkness over the whole land until three in the afternoon. (Matthew 27:45).

Prophecy forty-one: He would be buried in the tomb of a wealthy person.

Prophecy	Compliance
And His grave was assigned with wicked men, Yet He was with a rich man in His death (Isaiah 53:9).	When evening came, there arrived a rich man from Arimathea named Joseph, who had himself become a disciple of Jesus. [...] Joseph took the body, wrapped it in a clean linen shroud, and laid it in his own new tomb that he had hewn out of the rock. (Matthew 27:57-60).

I have presented the fulfillment of only forty-one prophecies—out of more than three hundred—delivered by eight different prophets: Moses, Isaiah, Zechariah, King David, King Solomon, Jeremiah, Amos, and Micah. These men lived for almost two millennia, from the 14th century BC to the 5th century AD. They spoke in different languages, lived in different geographical regions, and belonged to distinct historical contexts, yet each one offered specific and detailed descriptions concerning the coming of the Messiah.

Is this merely coincidence? Is it just luck that these prophecies were fulfilled with such precision?

FIFTH THESIS: THE PROBABILITY OF THE PROPHECIES BEING FULFILLED

I provide a straightforward explanation of the methods used to calculate probabilities in Appendix B, along with a brief discussion. For our purposes here, it is enough to understand that the probability of two independent events (m and n) occurring simultaneously is calculated as 1 divided by (m × n), or 1 / (m × n).

Let me illustrate this with a simple example. Suppose that one in ten men is over six feet tall, and one in one hundred men weighs over three hundred pounds. According to probability theory, only one in a thousand men (10 × 100 = 1,000) would be both over six feet tall and weigh more than 300 pounds.

To support this with a practical exercise, imagine randomly selecting 1,000 men and sorting them by height. Since one in ten men is over six feet tall, one hundred of them will meet this criterion, while the remaining nine hundred will fall below that height and therefore cannot possess both traits we are analyzing.

Now, focusing on the one hundred men who are over six feet tall, and applying the second condition—that only one in a hundred weighs over three hundred pounds—only one of these one hundred men will also meet the weight criterion. Thus, just one out of the original 1,000 will meet both conditions, confirming the result predicted by the formula.

For more than seventy-five years, *InterVarsity Christian Fellowship*[120] has established student groups at hundreds of universities worldwide, offering Bible study courses and promoting spiritual growth on college campuses. During the 1960s, the organization sponsored a remarkable five-year research exercise at Pasadena City College in California, USA.

The goal of this project was to examine the probability that a series of messianic prophecies could have been fulfilled by chance. Students were encouraged to use the most conservative and cautious methods available to make their estimates. For example, they considered questions such as: What is the likelihood that a random individual would enter the city of Jerusalem riding a donkey while claiming royal or divine authority?

More than six hundred students participated in this study over several semesters. They deliberated, analyzed probabilities, documented

---

[120]See www.intervarsity.org

their estimates, and presented their findings on a semester-by-semester basis[121].

What is the maximum number of individuals who could have entered Jerusalem on a donkey while claiming some form of authority? How common were donkeys in that period, and how many individuals would have had access to one? If ownership of such animals were typically limited to those with financial resources, how many people would qualify?

Now consider someone without wealth who still needed a donkey to fulfill such a symbolic act. That person would need to borrow the animal from a sympathetic, affluent acquaintance—someone willing to lend it for such a purpose. But how many individuals could realistically meet all these conditions?

By posing and discussing such specific, layered questions, the students participating in the study were able to reach a reasoned consensus on the statistical likelihood of any one person—at random—being able to fulfill this event exactly as described in prophecy.

The estimates I will use in the following analysis are direct results of this investigation. If you do not fully accept the numbers as they stand, you are welcome to adjust them to your own judgment. Even with reasonable modifications, the overall conclusion remains the same: the outcome is statistically astounding.

To support this argument, I will now present a detailed explanation of the exercise, using just eight of the forty-one prophecies I previously discussed.

Prophecy one: The Messiah was prophesied to be born in Bethlehem, to be the son of a virgin, and to be a descendant of King David—corresponding to prophecies two, six, and seven from the previous thesis.

The Gospels of Matthew and Luke both provide genealogical records of Jesus' lineage. Matthew 1:1–17 gives the genealogy through Joseph,

---

[121]See the book *Science Speaks: An Evaluation of Certain Christian Evidences* by Peter W. Stoner, M.Sc., who served as Chairman of the Department of Mathematics and Astronomy at Pasadena City College until 1953.

while Luke 3:23–38 traces it through Mary. The prophecy only required that the Messiah be a descendant of David, not through a specific parent, and both genealogies satisfy this requirement.

To understand the statistical significance of this fulfillment, we must consider the number of potential descendants from David to Jesus across twenty-five generations. Assuming each generation produced eight children, with a 50/50 male-to-female ratio, each generation would have four male children.

To calculate the number of potential male descendants, we raise 4 to the 25th power: 4^{25} = 1,125,899,906,842,624

This figure represents the total number of potential male-line descendants over twenty-five generations.

However, since the prophecy also indicated that the Messiah would be the firstborn, the total number must be reduced accordingly. That brings the figure down to 281,474,976,710,656.

Now, consider the second condition: the Messiah would be born in Bethlehem. At the time of Jesus' birth, Bethlehem was a small village with an estimated population of around three hundred people[122], while the world population is believed to have been approximately three hundred million[123]. This means that Bethlehem accounted for only one in every one million people on Earth.

So, to calculate the probability that someone who meets the first condition (a firstborn descendant of David) would also be born in Bethlehem, we divide: 1 / 281,474,976,710,656 × 1 / 1,000,000 ≈ 1 in 281,474,000,000 (or roughly 1 in 2.8×10^{11})

This gives a conservative estimate of the probability that one person, by random chance, could fulfill just three of the prophecies: being the firstborn, a descendant of King David, and born in Bethlehem.

Prophecy two: Another prophecy—prophecy sixteen from the previous thesis—stated that a messenger would precede the Messiah,

---

[122]See http://belenesdelmundo.com/wordpress/

[123]See https://magnet.xataka.com/un-mundo-fascinante/asi-ha-crecido-la-poblacion-humana-desde-el-ano-1-dc-hasta-la-actualidad

announcing His arrival. This messenger was identified as John the Baptist, who fulfilled this role in both message and character.

Now, consider the question: Of all the males who were born in Bethlehem, were firstborn, and descended from King David, how many could have had their coming announced in advance by a recognized prophet-like figure? The students involved in the study reasoned that such a messenger would have to be a unique individual, possessing the spiritual authority and qualities that characterized the prophets of antiquity.

They reached a conservative estimate, suggesting that only one in 1,000 individuals (or 1 in 10^3) could meet such a condition—having their appearance foretold by someone publicly recognized as a legitimate forerunner, such as John the Baptist.

Prophecy Three: Another prophecy—prophecy twenty from the previous thesis—foretold that the Messiah would be proclaimed king and would enter Jerusalem riding on a donkey.

The question then arises: How many of the men who were born in Bethlehem, were descendants of King David, firstborn, and whose coming had been announced by a prophetic messenger could also have fulfilled this specific event? While someone determined to "force" the prophecy might obtain a donkey and choose to enter the city through one of Jerusalem's gates, there is one crucial element he could not control: the reaction of the crowd. He could not manufacture the spontaneous proclamation of kingship by the people.

This element—being publicly recognized as a king during such an entrance—elevates the improbability of fulfilling this prophecy. The students conducting the analysis estimated that the chance of one individual meeting all these conditions and being hailed as king upon entering Jerusalem on a donkey was 1 in 10,000, or 1 in 10^4.

Prophecy four: Prophecy twenty-two from the previous thesis stated that the Messiah would be betrayed by one of His closest companions—a reference to the betrayal by the apostle Judas, one of Jesus' most trusted followers. This betrayal would lead to the wounds in His hands, also foretold in prophecy.

Unlike previous events, this prophecy is less causally connected to the others. It prompts a new question: What are the odds that a man—already meeting all the prior conditions—would be betrayed by a close friend, and that this betrayal would result in serious harm?

The students argued that such an event was highly uncommon. While betrayals do happen, they are far less likely to occur among individuals in close, trusted relationships, especially with consequences as severe as physical injury or death. This type of betrayal, particularly in a messianic context, would be especially rare.

Based on this reasoning, the students conservatively estimated the probability of such an occurrence as 1 in 1,000, or 1 in 10^3.

Prophecy Five: Prophecy twenty-three from the previous thesis states that the traitor would receive thirty pieces of silver in exchange for the betrayal.

In this case, the question is straightforward: Of the individuals who could have been betrayed by a close companion, how many would have been betrayed specifically for thirty pieces of silver? Not just any amount, but that exact figure.

The students agreed that such a precise detail—especially involving a specific and uncommon sum—would make the fulfillment of this prophecy extremely rare. It was not simply that a betrayal occurred, but that it was carried out for this exact price, adding another layer of specificity to the chain of fulfilled events.

As a result, they conservatively estimated the probability of this happening to be 1 in 10,000, or 1 in 10^4.

Prophecy Six: Prophecy twenty-four from the previous thesis foretold that the payment for the betrayal would be thrown into the temple and end up in its treasury. This prophecy is remarkably specific—it does not simply refer to the return of the money, but to the precise sequence of events involving the temple.

According to Matthew 27:3, Judas, feeling remorseful, attempted to return the thirty pieces of silver. The chief priests, however, refused to accept it. In response, Judas threw the coins into the temple before leaving. Later, the religious leaders used that money to purchase the potter's field, which became a burial place for foreigners who died in Jerusalem.

The students were instructed to calculate the probability of such a unique chain of events: that a man would betray a close friend for thirty pieces of silver, attempt to return it, have it rejected, then throw the money into the temple, where it would be retrieved by the priests and used to buy a cemetery.

After careful consideration, the students concluded that such an intricate and highly specific scenario would be extremely rare, and they assigned a conservative estimate of 1 in 100,000, or 1 in 10^5.

Prophecy Seven: Prophecy twenty-seven from the previous thesis stated that Jesus would remain silent during His trial, offering no defense for Himself even when facing the possibility of execution.

This detail stands out as deeply unusual. Most individuals, when falsely accused—especially in a life-threatening trial—would attempt to defend themselves, assert their innocence, or at least speak in their own favor. Remaining completely silent under such pressure would be extremely rare.

The students evaluating this prophecy were asked to consider how many men—who had already fulfilled all the previous criteria—would choose silence in a trial that could lead to their death. After examining historical behavior and likelihood, they estimated that the probability of such a response was 1 in 10,000, or 1 in 10^4.

Prophecy Eight: Prophecy thirty from the previous thesis foretold that the Messiah would be crucified—a prophecy made by King David, long before crucifixion was even used as a form of execution.

The question posed to the students was this: Since the time of King David, how many men have been crucified? Although crucifixion later became a common method of execution under Roman rule, it was eventually abolished centuries ago and is no longer practiced.

Taking the entire span of history into account, the students estimated that the number of individuals who were crucified since David's time would represent about 1 in 10,000 people, or 1 in 10^4.

Even if one were to disagree with the estimates made by the six hundred students involved in the study, the total probability of fulfilling these eight specific prophecies would remain astronomically low. As explained earlier in this thesis; to determine the probability of multiple independent events occurring simultaneously, one must multiply their individual probabilities.

So, let us do the calculation using the estimates provided by the students: $2.8 \times 10^5 \times 10^3 \times 10^4 \times 10^3 \times 10^4 \times 10^5 \times 10^4 \times 10^4 = 2.8 \times 10^{32}$

This means that only one in 10^{32} people could have fulfilled just these eight prophecies. Keep in mind, there are over three hundred prophecies concerning the Messiah, and I have only presented forty-one in the previous section. If we were to continue the exercise and include even a portion of the remaining thirty-three, the probability would become inconceivably smaller.

To better grasp the magnitude of 1 in 10^{32}, consider this: it is equivalent to 1 in 100,000,000,000,000,000,000,000,000,000,000. Now imagine we had that many silver dollar coins. If we used them to cover the entire surface of the Earth, they would form a layer approximately thirty-six meters (or about 120 feet) thick. Now, suppose we marked just one coin, blindfolded a person, and allowed them to walk anywhere on Earth, dig through the coins, and pick just one. The chance of that person selecting the marked coin on the first try would be the same as one man fulfilling all eight of those ancient prophecies by coincidence.

Yet Jesus of Nazareth did.

I have already demonstrated that these prophecies were written centuries before Jesus was born, based on manuscripts dating as far back as the eighth century BC. It is fair to assume that some prophecies—such as entering Jerusalem on a donkey—could theoretically have been intentionally fulfilled by someone attempting to present himself as the Messiah. It would not be difficult for a determined individual to acquire a donkey and ride it into the holy city.

But what about the rest? Could a person choose the place of his birth, or ensure He is a descendant of King David? Could He arrange to be betrayed by a close friend, or be crucified—a punishment no longer in use today? These are not controllable events.

So, how do we explain the fact that a single individual could fulfill so many specific and improbable predictions?

There are only two explanations. The first is that it was pure coincidence—that the prophets somehow guessed all these details without any divine insight, and that one man accidentally aligned with all of them. The second is that there was a higher intelligence—that God Himself revealed these future events to His prophets and orchestrated history to bring them to fulfillment in the person of Jesus Christ.

Professor Peter W. Stoner, former chair of the Department of Mathematics and Astronomy at Pasadena City College, built upon this analysis. He added eight more prophecies to the original eight and calculated[124] that the chance of one person fulfilling sixteen was 1 in 10^{53}. When he extended the analysis to forty-eight prophecies, the probability became 1 in 10^{181}.

To visualize this, remember that with eight fulfilled prophecies, one could cover the Earth in coins to a depth of one hundred and twenty feet. But at forty-eight, the layer of coins would extend beyond the sun.

With such incomprehensibly low odds, is it reasonable to believe that these prophecies were the result of chance, myth, or fabrication? If you, like me, find that impossible to accept, then only one conclusion remains:

These prophecies were inspired by God. This is the most compelling evidence for the true authorship of the Bible. Coincidence? Luck?

SIXTH THESIS: THE PROPHET DANIEL

The Old Testament is traditionally divided into four main sections: the Pentateuch, the wisdom books, the historical books, and the prophetic books. The prophetic books are further categorized into the major and minor prophets. These designations— "major" and "minor"—

---

[124]The calculation is found in his book *Science Speaks, an Evaluation of Certain Christian Evidences*.

do not refer to the importance of the prophets themselves, but rather to the length of their writings. Among the books classified under the major prophets is the Book of the Prophet Daniel.

Following the Babylonian invasion of Jerusalem in BC 587, led by King Nebuchadnezzar II, the Babylonian empire took many members of the Judean nobility into captivity. The king instructed Ashpenaz, the chief of his eunuchs, to select young Israelite men who were physically flawless, handsome, wise, well-educated, and intelligent—young men suitable to serve in the royal court.

These chosen individuals were to be nourished with food from the king's table and trained for three years in the literature and language of the Chaldeans, after which they would enter royal service. Daniel was among those selected. However, because of his faithfulness to his religious beliefs, he refused to eat the food and drink provided, as it violated Jewish dietary laws. Instead, he asked to be given only legumes and water for ten days. He proposed that, after the trial period, his physical condition be compared to that of the others who had eaten from the royal provisions.

At the end of the ten days, Daniel's health and appearance surpassed those of the other young men. This outcome gained him the respect and favor of his tutors, who subsequently committed themselves even more fully to his education. The Scriptures state that King Nebuchadnezzar found Daniel to be ten times wiser and more insightful than all the magicians and soothsayers in his entire kingdom.

Daniel's God-given ability to interpret dreams and visions soon elevated him to a position of great influence.

In chapters 10 and 11 of the Book of Daniel, the prophet receives a vision in which an angelic messenger reveals a detailed account of future events, spanning from the reign of Cyrus II the Great (BC 559–530) to that of Antiochus IV Epiphanes (BC 175–163), who ruled Persia and Syria, respectively. The vision is introduced in the third year of Cyrus's reign:

"In the third year of Cyrus king of Persia, a revelation was given to Daniel..." (Daniel 10:1). This placed the prophecy in the year BC 536[125].

As the vision unfolds, the angel declares:

> Now I shall tell you the truth about these things. Three more kings shall arise in Persia. Then a fourth will appear who will be far richer than all of them, and when he has enhanced his power through his wealth, he will mobilize the entire empire against the kingdom of Greece. (Daniel 11:2)

At the time this prophecy was given, Cyrus II was the reigning king of the Persian Empire, and Darius the Mede (also known as Gubaru) governed Babylon under Cyrus's authority.

The three kings mentioned by the angel are understood to be: Cambyses II (BC 530–522), the son of Cyrus II, Gautama (also called Pseudo-Smerdis or Bardiya) who ruled briefly in BC 522, and Darius I the Great (BC 522–486), who seized power after the assassination of the previous ruler. These three monarchs succeeded Cyrus II in direct succession.

In BC 486, Darius I the Great died at the age of sixty-three. His son, Xerxes I, also known as Xerxes the Great—or Ahasuerus in the Bible, a central figure in the Book of Esther—succeeded him. Xerxes I is the fourth king referenced by the angel in Daniel's prophecy. As foretold, Xerxes amassed great wealth and power and eventually launched the Second Persian War (also known as the Second Medical War) against the Greek alliance led by Sparta and Athens in the spring of BC 480.

At first, it appeared that Persia would win swiftly and decisively. However, despite initial successes, Xerxes' massive army retreated and returned to Asia. The Greek historian Herodotus[126], in his work *Histories*[127], claimed that Xerxes' army numbered over 1.7 million

---

[125]Daniel's mission began in BC 606, and the vision took place in the 70th year of his ministry.

[126]Herodotus of Halicarnassus was a Greek historian and geographer who lived between BC 484 and 425. He is traditionally regarded as the "Father of History" in the Western world. Herodotus was the first to compile a systematic and reasoned narrative of human events, seeking to explain not only what happened, but also why it happened.

[127]Volume VII, 60, 1.

soldiers—a figure now considered exaggerated by modern historians. Still, the number reflects the immense scale of Xerxes' mobilization and helps explain the final words of Daniel 11:2: "He will initiate all measures against the kingdom of Greece." The angel then continues in verse 3: "Then a mighty king will arise, who will rule with great power and do as he please."

This verse unmistakably refers to Alexander III of Macedonia, known to history as Alexander the Great.

Alexander the Great is regarded as one of the most formidable military conquerors of all time. He ascended to the throne of Macedonia in BC 336, at just twenty years old, following the assassination of his father, Philip II. Alexander had received extensive military training from his father and intellectual instruction from Aristotle[128], who influenced his education and worldview.

In BC 334, Alexander launched his ambitious military campaign against the Persian Empire. Over a span of just over ten years, he established one of the largest empires in the ancient world. His dominion stretched across the modern-day territories of Egypt, Israel, Lebanon, Jordan, Syria, Iraq, Iran, Afghanistan, Pakistan, Tajikistan, Turkey, Bulgaria, Greece, Serbia, and Croatia.

Alexander's conquest was not merely military—it was also cultural. With each new victory, his forces spread Greek language, philosophy, art, and governance across the region. This movement, strongly influenced by Aristotle's teachings, became known as Hellenization.

The angel continues:

> After he has arisen, his empire will be broken up and parceled out toward the four winds of heaven. It will not go to his descendants, nor will it have the power he exercised, because his empire will be uprooted and given to others. (Daniel 11:4).

---

[128]Aristotle (Stagira, BC 384 – Chalcis, 322) was a philosopher, polymath, and scientist born in the city of Stagira in northern Ancient Greece. Alongside Plato, he is considered one of the founding figures of Western philosophy. His ideas have had a profound and lasting impact on the intellectual history of the West for over two millennia.

The death of Alexander the Great in Babylon remains shrouded in mystery. He died in BC 323 at the age of thirty-three, without naming a clear successor. As a result, his vast empire was divided among his leading generals—known as the Diadochi.

Among these, Antigonus I Monophthalmos, Lysimachus of Thrace, Ptolemy I Soter, and Seleucus I Nicator emerged as the most powerful. Each took control of a region within the fragmented empire. Of these four, it was Ptolemy and Seleucus who played the most significant roles in the history of the Israelites, as their respective dynasties—the Ptolemies in Egypt and the Seleucids in Syria—fought for centuries over control of Judea and the surrounding region.

This prolonged struggle had profound effects on the Jewish people, who found themselves repeatedly caught in the middle of foreign domination and cultural pressure. *The Books of the Maccabees*, which are included in the Catholic Bible, recount the life of the Jews during this era of unending conflict.

The angel continues:

> The king of the South will become strong, but one of his commanders will become even stronger than he and will rule his own kingdom with great power. (Daniel 11:5).

The monarch being referred to is Ptolemy I Soter, who ruled Egypt until his death in BC 285. The general mentioned is Seleucus I Nicator, who, as foretold, annexed the territories of Media and Syria to Babylon after prolonged conflicts with his former companions-in-arms. These disputes among Alexander's successors led to the eventual division of the empire, and the emergence of the Seleucid and Ptolemaic dynasties, which would play a central role in the history of Israel and the surrounding region.

The angel continues:

> After some years, they will become allies. The daughter of the king of the South will go to the king of the North to make an alliance, but she will not retain her power, and he and his power[a] will not last. In those days she will be betrayed,

together with her royal escort and her father[b] and the one who
supported her. (Daniel 11:6).

After the death of Ptolemy I Soter, his son Ptolemy II Philadelphus
succeeded him and ruled Egypt until his death in BC 246. During his
reign, he ordered the translation of the Hebrew Scriptures into Greek, a
monumental work that became known as the Septuagint—a foundational
text for the Hellenistic Jewish world and early Christianity.

Meanwhile, in BC 281, Seleucus I Nicator died and was succeeded by
his son, Antiochus I Soter, who ruled the Seleucid Empire until BC 261.
After him, Antiochus II Theos, his son, ascended to the throne and
remained in power until his death in BC 246.

As foretold in Daniel's prophecy, a political marriage was arranged
to solidify peace between the Ptolemaic and Seleucid dynasties. In BC
261, Berenice Syra, daughter of Ptolemy II Philadelphus, was given in
marriage to Antiochus II Theos as part of a strategic alliance. To fulfill
the terms of this peace agreement, Antiochus was required to divorce his
first wife, Laodice I.

However, after the death of Ptolemy II, Antiochus II abandoned
Berenice and reconciled with Laodice. In an act of revenge, Laodice
ordered the murder of Berenice and Antiochus, an event that fulfilled the
prophecy recorded in the Book of Daniel.

The angel continues:

One from her family line will arise to take her place. He will
attack the forces of the king of the North and enter his fortress;
he will fight against them and be victorious. (Daniel 11:7)

From BC 246 to 222, the throne of Egypt was held by Ptolemy III
Euergetes, the brother of Berenice. At the same time, Syria was under the
rule of Seleucus II Callinicus, who remained in power until his death in
BC 225.

In fulfillment of his promise to avenge his sister's murder, Ptolemy
III declared war on Syria, initiating what became known as the Third
Syrian War. While he achieved some military success and inflicted
significant damage on the Seleucid territories, he did not secure a

decisive or lasting victory, and the war ended without fully accomplishing his objective of retribution.

The angel continues:

> He will also seize their gods, their metal images and their valuable articles of silver and gold and carry them off to Egypt. For some years he will leave the king of the North alone. Then the king of the North will invade the realm of the king of the South but will retreat to his own country. (Daniel 11:8-9).

During his campaign, Ptolemy III Euergetes managed to acquire an immense loot of 40,000 talents of silver and 2,500 sacred images of gods, many of which had originally been plundered from Egypt during the invasion by Cambyses II in BC 525. These religious artifacts had been taken to Persia, and Ptolemy's recovery of them during his invasion of Syria was seen as a remarkable national and religious triumph.

This significant accomplishment—restoring Egypt's stolen deities—earned him the title "Euergetes," meaning "benefactor."

The period of peace mentioned in Daniel's prophecy aligns perfectly with the peace treaty signed between Ptolemy III and Seleucus II Callinicus in BC 241, bringing a temporary end to the hostilities between Egypt and Syria.

However, Seleucus II later violated the treaty and attempted to invade Egypt, hoping to shift the balance of power in his favor. The effort failed, and he was forced to retreat, returning to his kingdom with less wealth than he had when he departed, just as the prophecy had foretold.

The angel continues:

> His sons will prepare for war and assemble a great army, which will sweep on like an irresistible flood and carry the battle as far as his fortress. Then the king of the South will march out in a rage and fight against the king of the North, who will raise a large army, but it will be defeated. (Daniel 11:10-11)

The sons of Seleucus II Callinicus took up the mantle of their father's ambitions for conquest. Upon Seleucus II's death, his eldest son, Seleucus III Ceraunus, ascended to the throne and ruled from BC 225 to

223. His reign was brief, and after his death, his younger brother, Antiochus III the Great, succeeded him.

One of Antiochus III's first military campaigns was directed against Ptolemy IV Philopator, the ruler of Egypt. The confrontation took place in the Lebanon region and resulted in a decisive defeat for Antiochus. However, despite this initial setback, Antiochus eventually managed to annex key strategic cities, including Tyre, Seleucia, and Ptolemais.

With these victories secured, Palestine became the next target. At that time, Palestine was under Egyptian control, and its Jewish population—caught in the middle—was forced to endure the clash of two powerful armies. The region's strategic importance and its vulnerable position made it a central battleground in the ongoing struggle between the Seleucid and Ptolemaic empires.

The angel continues:

> When the army is carried off, the king of the South will be filled with pride and will slaughter many thousands, yet he will not remain triumphant. For the king of the North will muster another army, larger than the first; and after several years, he will advance with a huge army fully equipped. (Daniel 11:12-13).

These formidable armies continued their struggle during what became known as the Fourth Syrian War. Antiochus III's forces, numbering approximately 62,000-foot soldiers, 6,000 cavalry, and 102 war elephants, advanced toward the gates of Egypt. In response, the Egyptian army, commanded by Ptolemy IV Philopator, assembled a phalanx[129] of 20,000 native troops, supported by Galatian and Thracian mercenaries and 73 African elephants.

The decisive confrontation took place at Rafah, located in the southern region of the Gaza Strip. It was there that Ptolemy's army

---

[129]The phalanx was a military formation developed in Ancient Greece and later adopted by various Mediterranean civilizations. It consisted of heavily armed infantry soldiers arranged in tightly packed rows, typically between eight and sixteen men deep. By extension, ancient authors often used the term "phalanx" to describe any army formation in which soldiers fought closely aligned in a unified front, following the model of the classical Greek phalanx.

achieved a significant victory, repelling the Seleucid advance and halting Antiochus' ambitions—for the time being.

Fourteen years later, Antiochus III returned, this time bearing plundered treasure, in fulfillment of the prophecy. His renewed strength and ambitions marked the continuation of the power struggle that had long defined the relationship between the Seleucid and Ptolemaic dynasties, with Palestine and its people caught at the heart of their ongoing conflict.

The angel continues:

> In those times many will rise against the king of the South. Those who are violent among your own people will rebel in fulfillment of the vision, but without success. Then the king of the North will come and build up siege ramps and will capture a fortified city. The forces of the South will be powerless to resist; even their best troops will not have the strength to stand. (Daniel 11:14-15)

Antiochus III appeared to have successfully restored the power and prestige of the Seleucid Empire in the East, earning him the title "the Great." During this time, Ptolemy V—only five years old—ascended to the throne of Egypt between BC 205 and 204, following the death of his parents. The power vacuum and the vulnerability of the young monarch presented an opportunity for Antiochus III to expand his influence.

Seizing the moment, Antiochus III entered into a secret agreement with Philip V of Macedonia to divide the Ptolemaic territories. According to the terms of this covert alliance, Antiochus would annex Cyprus and Egypt, while Philip V would gain control of regions near the Aegean Sea and Cyrene.

The phrase "many will rise against the king" from Daniel's prophecy is understood to refer to a specific group of Jews who, exhausted by the endless struggle between the Seleucid and Ptolemaic empires, chose to abandon the traditions of their ancestors. In doing so, they embraced the Hellenistic culture and pagan practices that Antiochus III promoted—

turning away from their religious identity in exchange for political or social advantage[130].

The angel continues:

> The invader will do as he pleases; no one will be able to stand against him. He will establish himself in the Beautiful Land and will have the power to destroy it. He will determine to come with the might of his entire kingdom and will make an alliance with the king of the South. And he will give him a daughter in marriage in order to overthrow the kingdom, but his plans will not succeed or help him. Then he will turn his attention to the coastlands and will take many of them, but a commander will put an end to his insolence and will turn his insolence back on him. After this, he will turn back toward the fortresses of his own country but will stumble and fall, to be seen no more. (Daniel 11:16-19)

In addition to conquering the "Beautiful Land"—Palestine, Antiochus III, who had earned also the title "the Great" for his military exploits, went on to plunder the cities he had captured during his campaigns. Interestingly, many of the local inhabitants celebrated the shift in power, hoping for stability under Seleucid rule.

To consolidate control over Egypt, Antiochus chose a diplomatic strategy. He negotiated a treaty with Ptolemy V Epiphanes, the young Egyptian pharaoh, and as part of the agreement, he offered his daughter, Cleopatra I Syra, in marriage. At the time of the pact, Ptolemy was just ten years old, and the marriage took place in BC 193, when he turned fourteen.

However, the strategy failed for Antiochus. His daughter refused to cooperate with her father's political aims and sided with her husband, undermining Antiochus's intentions to influence Egypt through her.

Turning his ambitions elsewhere, Antiochus launched military campaigns across the Aegean islands, where he experienced some initial victories. But his success was short-lived. In BC 190, he suffered a decisive defeat at the Battle of Magnesia, at the hands of Publius Cornelius Scipio

---

[130]Palestine remained under Ptolemaic control following the time of Alexander the Great—an era during which many began to abandon their traditions and observance of the Law, as described in the biblical books of the Maccabees.

Africanus, the famed Roman general. As the prophecy had indicated, this defeat marked the beginning of the end for Antiochus III.

In the aftermath, the Roman government forced Antiochus to surrender much of his territory and pay massive tribute. He returned to his homeland following the signing of an armistice, in which he pledged not to wage war against any Roman province or its allies.

Antiochus III met a dishonorable end. In BC 187, he was assassinated while attempting to loot treasures from a temple, a desperate act that closed the final chapter of his tumultuous reign.

The angel continues:

> His successor will send out a tax collector to maintain the royal splendor. In a few years, however, he will be destroyed, yet not in anger or in battle. (Daniel 11:20)

The successor of Antiochus III was his son, Seleucus IV Philopator, sometimes also referred to as "the Great." Seleucus ruled for twelve years and faced significant financial difficulties throughout his reign. These challenges were due to the heavy tribute payments owed to Rome; a burden inherited from the defeat and treaty conditions imposed on his father following the Battle of Magnesia.

To raise funds and meet these obligations, Seleucus IV dispatched his official Heliodorus to Jerusalem in BC 176 to seize the treasures of the Temple, an event recorded in 2 Maccabees 3. This act was not only sacrilegious but also deeply provocative to the Jewish people.

As the prophecy foretold, the events took a dramatic turn: upon returning, Heliodorus assassinated Seleucus IV, bringing an abrupt end to his reign.

The angel continues:

> He will be succeeded by a contemptible person who has not been given the honor of royalty. He will invade the kingdom when its people feel secure, and he will seize it through intrigue. (Daniel 11:21)

Demetrius I Soter, the son of Seleucus IV, was the rightful heir to the throne following his father's death. However, due to the debt obligations incurred by his grandfather, Antiochus III, Demetrius was held in Rome as a hostage, serving as a security guarantee for the ongoing tribute payments to the Roman Empire.

With Demetrius detained, the throne was assumed by Seleucus IV's brother, Antiochus IV Epiphanes. Antiochus did not acquire power through legitimate succession but rather through political maneuvering and strategic deception. His manipulative ascent to power—taking advantage of his nephew's absence—aligns precisely with the deceptive schemes described in Daniel's prophecy.

The angel continues:

> Then an overwhelming army will be swept away before him; both it and a prince of the covenant will be destroyed. After coming to an agreement with him, he will act deceitfully, and with only a few people he will rise to power. When the richest provinces feel secure, he will invade them and will achieve what neither his fathers nor his forefathers did. He will distribute plunder, loot and wealth among his followers. He will plot the overthrow of fortresses—but only for a time. "With a large army he will stir up his strength and courage against the king of the South. The king of the South will wage war with a large and very powerful army, but he will not be able to stand because of the plots devised against him. Those who eat from the king's provisions will try to destroy him; his army will be swept away, and many will fall in battle. (Daniel 11:22-27).

Antiochus IV Epiphanes, often referred to as the "ruthless king," waged wars of such intensity and scale that they rendered the conflicts of his ancestors almost insignificant by comparison. One of his calculated political moves was to extend a pact of friendship to his brother-in-law, the Egyptian pharaoh. However, this alliance was short-lived. Antiochus soon violated the pact, launching an invasion that allowed him to conquer nearly all of Egypt, except for its capital, Alexandria.

To avoid provoking Rome, Antiochus chose not to assume direct control over the Egyptian throne. Instead, he restored King Ptolemy VIII Physcon to the throne, in line with the agreement he had made with his nephew Ptolemy VI Euergetes. Yet this restoration was symbolic—

Ptolemy VI returned to power only as a puppet, firmly under the control of his Seleucid captor.

The angel continues:

> The king of the North will return to his own country with great wealth, but his heart will be set against the holy covenant. He will take action against it and then return to his own country. (Daniel 11:28)

The Romans, under the command of Consul Gaius Popilius Lenas, intervened and forced Antiochus IV to withdraw from Egypt. The famous encounter between Popilius and Antiochus included the consul drawing a circle in the sand around the Seleucid king, demanding that he decide before stepping out—a clear demonstration of Roman authority.

Antiochus had no choice but to comply. He returned to Syria, abandoning his ambitions in Egypt. However, he did not leave empty-handed. Antiochus brought back great wealth, not only from Egypt but also from his plundering of Jerusalem during his campaign.

The angel continues:

> At the appointed time he will invade the South again, but this time the outcome will be different from what it was before. Ships of the western coastlands will oppose him, and he will lose heart. Then he will turn back and vent his fury against the holy covenant. He will return and show favor to those who forsake the holy covenant. (Daniel 11:29-30)

In BC 168, after losing control of his puppet ruler, Ptolemy VIII Euergetes—the brother of Ptolemy VI—to the Alexandrian populace, Antiochus IV Epiphanes resolved to launch a new assault on Egypt. He briefly succeeded in occupying Cyprus during this campaign. However, the Romans once again intervened, forcing him to withdraw from all occupied territories.

Frustrated and humiliated, Antiochus turned his fury toward the Jews in the Holy Land during his return. On December 16 in BC 167, in a blatant act of religious provocation and oppression, he ordered the construction of an altar to Zeus in the very spot where the altar of burnt offerings once stood in the Jerusalem Temple. To further desecrate the sacred space and eradicate Jewish religious practices, Antiochus offered a pig—an unclean animal in Jewish law—as a sacrifice to his god.

These shocking events marked the beginning of intense persecution and are vividly chronicled in the *First Book of Maccabees*:

> Then the king issued an edict to his whole kingdom that all of his subjects should become a united people, with each nation abandoning its particular customs. All the Gentiles accepted the decree of the king, and many among the Israelites adopted his religion, sacrificing to idols and profaning the Sabbath. The king also sent messengers to Jerusalem and the cities of Judah with edicts commanding them to adopt practices that were foreign to their country: to prohibit holocausts, sacrifices, and libations in the sanctuary, to profane the Sabbaths and feast days, to defile the temple and its priests, to build altars, temples, and shrines for idols, to sacrifice swine and other unclean beasts, to leave their sons uncircumcised, and to allow themselves to be defiled with every kind of impurity and abomination, so that they would forget the law and change all their observances. Anyone who refused to obey the command of the king was to be put to death. [...] On the fifteenth day of the month Chislev, in the year one hundred and forty-five, the king erected upon the altar of holocausts the abomination that causes desolation, and pagan altars were built in the surrounding towns of Judah. Incense was offered at the doors of the houses and in the streets. Any scrolls of the law that were found were torn to pieces and destroyed by fire. If any people were discovered in possession of a book of the covenant or acting in conformity with the law, they were condemned to death by the decree of the king. Month after month these wicked people used their power against any loyal Israelite found in the towns. (1 Maccabees 1:41-58)

The angel continues in his revelation of future events to Daniel:

> His armed forces will rise up to desecrate the temple fortress and will abolish the daily sacrifice. Then they will set up the abomination that causes desolation. With flattery he will corrupt those who have violated the covenant, but the people who know their God will firmly resist him. "Those who are wise will instruct many, though for a time they will fall by the sword or be burned or captured or plundered. (Daniel 11:31-33)

Returning to the *First Book of Maccabees*, we witness the complete fulfillment of the prophetic episode that marked the beginning of the

Maccabean War. The first to rebel against Antiochus IV's edict was an elderly priest named Mattathias, a devout man, and the father of five sons. His righteous anger at the king's desecration of the Temple and the enforcement of pagan worship compelled him to act.

In a dramatic act of defiance, Mattathias killed the king's emissary, who had come to enforce the new law, along with those complicit in erecting the pagan altar. He then fled to the mountains with his sons, where they began to organize a guerrilla resistance movement to fight against the Seleucid forces.

Shortly after initiating the rebellion, Mattathias died, but the leadership of the resistance passed to his son Judas, who would later earn the title "Maccabeus" (meaning "the Hammer"). Under Judas's command, the Maccabean militia grew in strength and resolve.

In December of BC 164, the Maccabees successfully recaptured Jerusalem, an event recorded in 1 Maccabees chapters 2–4.

The angel continues:

> When they fall, they will receive a little help, and many who are not sincere will join them. Some of the wise will stumble, so that they may be refined, purified and made spotless until the time of the end, for it will still come at the appointed time. (Daniel 11:34-35).

During the period of resistance, many individuals joined the Maccabean guerrilla movement, not out of religious conviction or a desire to preserve Judaism, but simply as a means of survival. Faced with oppression and the threat of death, they aligned themselves with the rebels to escape persecution.

However, this prolonged conflict served a greater purpose beyond military resistance. It became a time of refinement and purification for the nation of Israel. The hardships and sacrifices exposed true loyalty, separating those who genuinely upheld the faith of their ancestors from those who had merely sought refuge in the movement.

This period of trial and testing was not only historical—it was also prophetic. The prophet Zechariah had foretold such a time:

Throughout the land, says the Lord, two-thirds in it will be cut off and perish, and one-third will be left. I will put that one-third through fire, and I will refine them as silver is refined, and I will test them as gold is tested. They will call on my name and I will hear them. I will say, "These are my people," and they will say, "The Lord is our God." (Zechariah 13:8-9)

Verses 36–45 of the Book of Daniel continue to describe the reign of Antiochus IV Epiphanes, focusing on the atrocities he would commit against the Jewish people. While many of these prophetic details align closely with the known historical account of his life and actions, there are a few elements that are difficult to situate precisely within the narrative of this brutal ruler.

One such detail is the location of his death. The prophecy appears to imply that he would die near Jerusalem, yet historical records confirm that Antiochus IV died in Persia. Despite this geographical discrepancy, the prophecy accurately reflects the nature of his death—a sudden and excruciating demise, filled with humiliation and physical suffering.

The Second Book of Maccabees offers a vivid description of his end:

About that time it so happened that Antiochus was leading an ignominious retreat from the region of Persia. He had entered the city called Persepolis and attempted to plunder the temple and gain control of the city. However, the people immediately rose up in armed defense and repulsed Antiochus and his men, with the result that Antiochus was put to flight by the inhabitants and forced into a humiliating retreat. On his arrival in Ecbatana, he learned what had happened to Nicanor and to the forces of Timothy. Bursting with anger, he devised a plan to make the Jews suffer for the injury inflicted by those who had put him to flight. Therefore, he ordered his charioteer to drive without stopping until he completed his journey. However, the judgment of Heaven rode with him, since in his arrogance he declared, "Once I arrive in Jerusalem, I will turn it into a mass graveyard for Jews." And so the all-seeing Lord, the God of Israel, struck him with an unseen but incurable blow. Hardly had he spoken those words when he was seized with excruciating pains in his bowels and acute internal torment— an entirely suitable punishment for one who had inflicted many barbarous torments on the bowels of others. Nevertheless, he did not in the least diminish his insolent behavior. More arrogant than ever and breathing fire in his rage against the Jews, he gave orders to drive even faster. As a result, he was hurled from the lurching chariot, and the fall was so violent that every part of his body was racked with pain. Thus he who only a short time before had in his superhuman arrogance believed that he could command the waves of the sea, and who imagined that he could weigh high mountains on a scale, was thrown down to the ground and had to be carried in a litter, clearly manifesting to all the power of God. The body of this ungodly man swarmed with worms, and while he was still alive suffering agonizing torments, his flesh rotted away, so that the entire army was sickened by the stench of his decay. Only a short time before, he had thought that he could touch the stars of heaven. Now no one could even bring himself to transport the man because of his intolerable stench. Ultimately, broken in spirit, he began to lose his excessive arrogance and to come to his senses under the scourge of God, for he was racked with incessant pain. When he no longer could endure his own stench, he exclaimed: "It is right to be subject to God. Mere mortals should never believe that they are equal to God." Then this vile wretch made a vow to the Lord, who would no longer have mercy on him, that he would publicly declare to be free the holy city toward which he had been hurrying to level it to the ground and transform it into a mass graveyard; that the Jews, whom he had not deemed to be worthy of burial but fit only to be thrown out with their children and eaten by wild animals and birds, would all be granted equality with the citizens of Athens; that the holy temple that he had previously plundered, he would now adorn with the finest offerings, replace all the sacred vessels many times over, and provide from his own revenues the expenses incurred for the sacrifices. In addition to all this, he would become a Jew himself and would visit every inhabited place to proclaim the glory of God. [...]

And so this murderer and blasphemer, after enduring agonizing sufferings to match those he had inflicted on others, died a wretched death in the mountains of a foreign land. His close friend Philip brought back the body. Then, fearing the son of Antiochus, he withdrew into Egypt, to the court of Ptolemy Philometor. (2 Maccabees 9)

All the historical facts discussed in this section of the chapter can be independently verified through reputable historical sources. As a result, you can be confident that the prophecies recorded in the Book of Daniel were fulfilled with an extraordinary level of accuracy and detail—a level that is difficult, if not impossible, to explain apart from divine revelation.

The Old Testament is filled with clear and precise prophecies, and it is this very accuracy that grants the authors the rightful title of "prophet." The same prophetic authority that predicted the rise and fall of kings and empires also proclaimed the coming of the Messiah. These messianic prophecies were delivered with just as much clarity, specificity, and divine weight.

So how can we explain this? Coincidence? Luck?

CONCLUSION

Throughout the many years I have dedicated to lecturing on our faith, I have encountered an astonishing number of falsehoods and legends surrounding biblical topics. Sadly, many believers remain unfamiliar with the origins and historical preservation of the Holy Bible. They often overlook the fact that the Bible is the most thoroughly documented work of antiquity—far surpassing all other texts of its era in terms of manuscript evidence and historical support.

Based on the information I have presented in this chapter, those who hold a Bible in their hands can be assured that this sacred text carries the same message that its original authors intended, without corruption or manipulation over time.

Thanks to the preservation of thousands of ancient manuscripts—now housed in museums and libraries around the world—the public could examine and compare these ancient sources. Such comparisons consistently affirm the faithfulness of the biblical message, demonstrating that despite the passing of centuries, its content has been reliably transmitted.

I also presented evidence regarding the approximate period in which the biblical manuscripts are believed to have originated, based on the consensus of leading scholars and experts. By using this information, we can identify the timeframe in which the author lived. If that author predicted an event that later came to pass, and the writing is demonstrably older than the event itself, then we possess clear evidence that the writer was a true prophet.

The dating of the prophecy is crucial. It allows us to confirm that the prophecy was indeed written before the event occurred. While we may not always have the precise date the prophecy was delivered, having a reliable timeframe is sufficient. The predictions examined in this chapter describe events that took place centuries after they were made. The Book of Daniel, for example, is preserved in manuscripts from his era, confirming that his prophecies predated the events they so accurately described.

Of course, people have been making predictions about the future since before the time of Moses, and many continue to do so in the world today. However, simply attempting to foresee the future does not make someone a prophet.

As shown throughout this chapter, a true prophecy must meet two essential criteria. First, the prophecy must be revealed—a divine disclosure that reflects the prophet's close relationship with God. Second, the prophecy must be fulfilled—the predicted event must actually come to pass.

Those who personally encountered the prophets in biblical times could attest to the first requirement, often identifying divine calling through miracles[131] or the prophet's own sacrifice or suffering[132]. However, they could not yet verify the second requirement, as the events were still in the future.

Over time, as the events unfolded exactly as foretold, the people of Israel and later generations came to recognize the authenticity of the prophets' words. Their prophecies were then preserved in Scripture—not only as records of what was said, but also as testimonies of fulfilled divine revelation.

We are currently witnessing the fulfillment of hundreds of prophecies, many of which can now be clearly recognized and dated with precision, even though they were predicted centuries in advance. The accuracy of the prophecies revealed by Daniel in chapter eleven is far beyond the reach of human speculation or imagination.

---

[131]See 1 Kings 17:17–24, Exodus 14:21–31, Numbers 20:7–11, Numbers 22:21–35, Joshua 10:12–14, 1 Samuel 12:18, 2 Kings 4:2–7, Daniel 6:16–23, Jonah 2:1–10, among others.

[132]According to tradition, the prophet Isaiah was killed by King Manasseh. Jewish tradition also holds that the prophets Ezekiel and Jeremiah died as martyrs.

It is essential to remember that Daniel's prophecies encompass some of the most significant events across a 400-year span of history. This prophetic timeline begins during the reign of Cyrus II the Great (BC 559–530) in Persia and concludes during the reign of Antiochus IV Epiphanes (BC 175–163) in Syria.

Ask yourself: Is it feasible for anyone to construct a detailed narrative, filled with marriages, conquests, defeats, political alliances, royal successions, betrayals, inheritances, wars, exiles, heroes, villains, victors, and victims, hundreds of years before those events unfolded?

Is this not conclusive evidence that Daniel's words reflect the voice of the One who owns and commands history? Could Daniel himself have been unaware of God's calling and divine selection? And if this is true of Daniel, does it not raise the same question for every other prophet who wrote with such authority and accuracy?

This chapter has demonstrated a variety of fulfilled prophecies—some concerning blessings and restoration, others foretelling judgment and suffering. Yet the greatest prophecies, the most profound and consequential, are those which declared that God would become man—that He would be born of a virgin, live among us, and through His life, death, and resurrection, transform the course of human history.

These messianic prophecies are the most significant, not because they were the most dramatic, but because of their universal impact.

God had made a most significant promise to Abram:

> The Lord said to Abram, "Leave your country, your people, and the house of your father, and go to the land to which I will lead you. "I will make of you a great people and I will bless you. I will make your name great and it will become a blessing. I will bless those who bless you and curse those who curse you. And through you all the nations on the earth shall be blessed." (Genesis 12:1-3)

"God's chosen people" were the descendants of Abram, and they carried this divine promise in their hearts and minds throughout their lives. It was a truth instilled in them from childhood, and they held on to it with unshakable hope, even as they approached death—the conviction that the promise would soon be fulfilled. God has chosen us to be His people. Could there be any greater hope? What greater future could exist for a people chosen by the Creator from among nations, entrusted with the purpose of blessing all the families of the earth?

And yet, the promise came with a single expectation: fidelity. God required faithfulness in return for the covenant established through their encounter with Him. Despite witnessing miracles and undeniable acts of divine power, the people repeatedly failed to uphold their part of the covenant. As a result, they were subjected to a long sequence of captivities and foreign domination—first by the Egyptians, then the Babylonians, Medes, Persians, Greeks, and finally the Romans.

There were brief periods of greatness, particularly during the reign of King David, when the people believed they might finally embrace the long-awaited promise. But once again, human passions prevailed, and they turned their backs on the Lord. Nevertheless, hope never vanished. When the prophets began to proclaim that God would send His Son to restore Israel, Jewish hearts were filled with expectation (Luke 2:25). The coming of the Messiah became their deepest longing—a Redeemer who would bring freedom, prosperity, and glory unmatched by any other nation.

The prophets did not speak in vague or cryptic terms. They gave precise, detailed information that would enable clear and unmistakable identification of the Messiah. I compiled a list of forty-one prophecies that are the most straightforward to recognize, yet the total number exceeds three hundred.

So how can we explain this?

How could so many details—given by dozens of individuals, living in different eras, spread across various regions, and with no communication among them—align so perfectly to identify one person?

According to my analysis, the probability of a single individual fulfilling these prophecies is 1 in 10^{181}. And yet, it happened.

To suggest this is a mere coincidence is to believe that such an outcome could occur by chance—by an unimaginable series of random alignments. But this number does not point to luck. It serves as clear, objective, and compelling evidence that the prophets spoke because God revealed it to them.

Those who are today referred to as "seers" or clairvoyants, celebrated for their supposed ability to foresee the future, are in most cases simply engaging in a form of statistical analysis. They gather as much data as possible on a given subject and then make predictions based on trends and probabilities. Whether it is forecasting the outcome of the next World Cup, the results of a presidential election, or a potential plane crash in Europe, their predictions can often be explained through probability theory combined with extensive research and publicly available information.

If their predictions prove wrong, the consequences are minimal—at worst, they lose followers or public interest and may need to find another platform or career. In contrast, the stakes for ancient prophets were immeasurably higher.

In biblical times, claiming to speak for God was not taken lightly. To falsely present oneself as having direct communication with the Creator was considered a grave offense. The penalty for doing so was not ridicule or a drop in popularity—it was often banishment or even death. The Old Testament contains some of the strongest warnings and condemnations against false prophets:

> This word of the Lord came to me: Son of man, prophesy against the prophets of Israel who are now prophesying. Say to those whose prophesies are formulated in their own minds: Hear the word of the Lord. Thus says the Lord God: Disaster will engulf those foolish prophets who follow thoughts that are fabricated in their own imaginations and have received no visions. Your prophets, O Israel, are like jackals foraging among ruins. They have not bothered to reinforce the breaches in the walls of the house of Israel so that it may stand firm in battle on the day of the Lord. The visions they saw were false, and their divinations were baseless. They assert: "Thus says the Lord," despite the fact that the Lord did not send them, and then they expect their words to be proved true. Have you not seen false visions or uttered lying divinations when you have asserted, "Thus says the Lord," even though I have not said any such thing? Therefore, thus says the Lord God: Because you have spoken untruths and proclaimed false predictions, I have now set myself in opposition

to you, says the Lord God. My hand will be raised against those prophets whose visions are baseless and whose divinations are clearly false. They will not be granted any position in the council of my people, nor will their names be enrolled in the register of the house of Israel, nor will they be permitted to set foot in the land of Israel. Then you will know that I am the Lord. Because they lead my people astray, crying aloud, "Peace!" when there is no peace, and because, when the people were repairing a flimsy wall, these prophets concealed its flaws by smearing whitewash on it, say to those who covered it with whitewash that it will collapse, for I will cause rain to fall in torrents, and I will send hailstones hurtling down and unleash a wind of gale force. When the wall collapses that you have smeared with whitewash and it falls to the ground so that its foundations will be laid bare, you will be destroyed along with it, and thus you will know that I am the Lord. Therefore, thus says the Lord God: I intend to unleash a violent storm wind in my rage, torrential rain in my anger, and hailstones in my fury, and I will shatter the wall that you smeared with whitewash and knock it to the ground and lay bare its foundations. It will fall, and you will perish beneath it. Then you will know that I am the Lord. When I have vented my fury upon the wall and upon those who smeared it with whitewash, I will say to you, "The wall is gone, and so are those who smeared it— the prophets of Israel who prophesied about Jerusalem and envisioned peace for it when there was no peace," says the Lord God. (Ezekiel 13:1-16)

The indisputable confirmation that God has maintained communication with us—His children— lies in the complete fulfillment of the many prophecies that foretold detailed aspects of the Messiah's life. Through these revelations, God not only spoke to us, but also gave us the assurance that those whom He chose as His messengers were genuine, and that His message is true.

Can we honestly believe that a prophet who accurately described the coming of the Messiah—often centuries in advance—could have somehow lied about everything else? Is it not compelling evidence that these exceptional individuals were indeed inspired by God?

Why, then, would we assume that our Heavenly Father desired to communicate only during the prophetic era, only to fall silent thereafter? As I stated in the introduction to this chapter, God's Word is unchanging, just as He Himself is immutable. What He revealed through the prophets remains as relevant today as it was in their time.

Therefore, it is entirely correct to affirm that God continues to communicate with us—not only through the beauty of Creation, and the deep emotions written on the human heart, but also, and most explicitly, through the living and enduring Word found in the Bible.

God chose Moses as one of the prophets who shared the closest relationship with Him. On numerous occasions, they engaged in direct communion that lasted for days, such as during the time Moses received the Ten Commandments: "Moses remained with the Lord for forty days and forty nights" (Exodus 34:28).

During these extended encounters, Moses had the unique opportunity to speak with God on a wide range of subjects—including the origin of the universe. Like many of us, Moses was curious. He wanted to understand how everything began, where it came from, and how it all came into being.

It is important to emphasize that, while we now live in an age of scientific advancement, where many pieces of the Creation puzzle are slowly being assembled, this knowledge is very recent. Just a hundred years ago, much of what we now understand was still a mystery. Even more so in the time when Genesis was written, approximately 3,700 years ago.

And yet, in Genesis, Moses described Creation in a way that remarkably aligns with the modern scientific understanding of the universe. How can this be explained? How could someone from an ancient, pre-scientific era record an account that so closely matches what science confirms today?

Of course, we must account for the non-technical language used in Genesis—understandable and appropriate for a message intended for all people across all generations. But consider the essence of what Moses recorded: that the universe had a beginning, that light emerged, that life came from matter, that water was the starting point of existence, and so on. These concepts align with scientific theories such as the Big Bang and the evolution of life from primordial elements.

Is this not powerful evidence of Moses' extraordinary closeness to the Creator? Could such accuracy be possible without divine guidance?

The information presented in Genesis stands as an empirical indicator of the Creator's communication with humanity. Once again, to suggest that the author of this narrative was simply lucky in correctly describing each major event of Creation—and doing so with such remarkable detail and accuracy—is to stretch the limits of the law of probability beyond reason.

Other religions, like those I referenced earlier in this argument, chose more poetic and symbolic paths in their attempts to answer the great question of the universe's origin. Their authors, drawing from myth and imagination, crafted stories that sought meaning but lacked detail— especially when compared to the modern scientific understanding of the cosmos.

In contrast, it is our Bible that presents a narrative with specific elements that, despite being written in ancient times and in non-technical language, remarkably align with the current scientific account of Creation. This astonishing convergence defies purely human reasoning or historical coincidence. It is further evidence that the true author of Genesis is none other than the Creator Himself—the One who possessed perfect knowledge of the events described.

Only the Owner of Creation could have provided such a narrative. He alone had access to the full truth behind the formation of the universe, and He chose to reveal it through divine revelation to His servant.

This reality offers a compelling answer to a set of common questions posed by many deists and skeptics: What makes us think we follow the true God? How can we be sure we are not worshiping the wrong deity? Why not the god of Hinduism, or another ancient faith?

If those gods were the true creators, then their sacred texts would contain creation accounts that align with observable reality. But as we know, they do not. Their narratives, while rich in metaphor, diverge sharply from scientific knowledge.

The Bible has never claimed to be a textbook of science, geography, or astronomy. Yet it is impossible to ignore the fact that it contains information in these areas—surprising, accurate information that was entirely unknown at the time it was written. The biblical authors referred to a wide range of facts that humanity would not come to understand until just the last few centuries.

Can the references I have presented in this chapter truly be dismissed as mere poetry? I acknowledge that it's possible some of these statements—such as the mention of the vast number of stars, their differences, the Earth floating in space, its roundness, the water cycle, or even the first and second laws of thermodynamics—were written using figures of speech, like the ones discussed earlier in this work. Perhaps the authors were indeed using poetic language.

But then the real question arises: Why did these "poetic" descriptions turn out to be scientifically accurate—validated thousands of years later? And why do we not find the same level of clarity or truth in the sacred texts of other religions?

Beneath all the evidence presented in this chapter lies a single unifying conclusion: The Bible could not have been written by human intellect alone. Dozens of authors, most of whom never met, living thousands of kilometers apart, across different cultures, eras, and empires, speaking different languages, and coming from vastly different backgrounds—from slaves to kings, from murderers to generals—produced seventy-three books that are remarkably consistent, theologically unified, and free from contradiction.

In addressing the question of whether God communicates with us, the reader should come away with a deep sense of peace and certainty: He has indeed spoken. Through the Bible, He established a secure, enduring communication bridge with us—His children—as we wait for the day, we are united with Him.

Does God communicate with us? There is no doubt about it.

He has, and He still does.

CAN WE TRUST THAT COMMUNICATION?

As they approached the village to which they were going, he acted as though he would be going further. However, they urged him strongly, "Stay with us, for it is nearly evening and the day is almost over." And so he went in to stay with them. When he was at table with them, he took bread, blessed and broke it, and gave it to them. Then their eyes were opened and they recognized him, but he vanished from their sight. They said to each other, "Were not our hearts burning within us while he spoke to us on the road and opened the Scriptures to us?" They set out immediately and returned to Jerusalem, where they found gathered together the Eleven and their companions who were saying, "The Lord has truly been raised, and he has appeared to Simon!" Then the two described what had happened on their journey and how he had made himself known to them in the breaking of the bread.

LUKE 24:28-35

In the mid-1970s, Uri Geller—a then-famous psychic of Israeli origin—visited Colombia to demonstrate his mental powers on national television. I still remember how my entire family gathered around the television, captivated by the anticipation of seeing him bend a spoon using nothing but his mind. With a close-up shot trained on his hands, he rubbed the utensil between his thumb and forefinger. There was no trickery, no sleight of hand—everything unfolded before our eyes. I watched, astonished, as the metal appeared to soften and twist as if it were melting.

The highlight of the broadcast came when Geller claimed he could repair damaged watches belonging to viewers at home using only his mind. My brother sprang up to find one of our broken old watches and followed Geller's instructions to the letter. But the clock hands remained still. We figured the watch might have needed to be closer to the television—to better absorb the mental energy he was supposedly transmitting. The next day at school, several classmates excitedly claimed that their long-dead clocks had miraculously come back to life.

For several years, I genuinely believed that powers like those Uri Geller displayed were real. How could I not? I had witnessed them live, right before my eyes! The enchantment, however, began to fade in the late 1970s when James Randi[133]—a renowned illusionist and escape artist, well known for his appearances on the television show *Wonderama*—publicly accused Geller of being a fraud. According to Randi, Geller used standard magician's tricks and passed them off as manifestations of psychic power.

Randi repeatedly challenged Geller to demonstrate his abilities under controlled conditions, but Geller never accepted. Undeterred, Randi detailed his accusations in the book *The Magic of Uri Geller*, where he carefully explained the sleight-of-hand techniques magicians used to replicate every one of Geller's feats of mental power.

Building on his skepticism—and following a contentious radio debate with a parapsychologist—Randi established what became known as the *Million Dollar Paranormal Challenge*[134] in 1964. The premise was simple: anyone who could demonstrate a supernatural or paranormal ability under controlled conditions would receive a cash reward. The challenge originally offered $10,000, but by the time it ended in 2015, the award had risen to $1 million. Over the decades, around thousand people attempted the challenge, but not even one succeeded in meeting the foundation's rigorous conditions.

---

[133]Randall James Hamilton Zwinge (born August 7, 1928, in Toronto), better known as James Randi, was a Canadian illusionist, writer, and skeptic. He became widely known in the United States media for exposing frauds related to parapsychology, homeopathy, and other pseudosciences. With nearly fifty years of experience as an illusionist, Randi possessed exceptional skill in identifying the deceptive techniques used by individuals claiming to have supernatural powers.

[134]See https://web.randi.org/

Today, more than a hundred organizations[135] across the globe offer similar rewards for verified demonstrations of paranormal powers. To date, all the prizes remain unclaimed.

Since ancient times, humanity has marveled at astonishing feats of magic—illusions so convincing they could easily be mistaken for supernatural powers. One of the earliest recorded examples comes from the biblical account of Moses and Aaron before the Pharaoh of Egypt (Exodus 7–12). When they demanded the release of the Jewish people, God instructed them that if Pharaoh asked for a sign, they should cast down Aaron's staff, which would transform into a serpent. They obeyed, and the staff became a snake.

Pharaoh, unimpressed, summoned his own wise men and magicians. According to the scriptures, they too cast down their staff, which likewise turned into serpents—although Aaron's snake devoured the others. This begs the question: did Pharaoh's magicians possess true supernatural powers? Likely, they did not. They were skilled illusionists who understood the craft of deception.

Egyptian sorcerers were known for their mastery in snake charming. One well-documented technique involved pressing on a snake's neck to induce a trance-like state, causing it to stiffen and appear lifeless—like a wooden rod. Concealed within their garments, such snakes could be dramatically revealed in a way that mimicked a miraculous transformation. The renowned magician Walter B. Gibson[136] details this very trick in his book *Secrets of Magic*, outlining how such sleight of hand was used to amaze audiences and simulate the impossible.

After Pharaoh's initial refusal, Moses and Aaron returned to make their request once more. Again, they met with rejection. In response, Aaron stretched out his staff over the waters of the Nile, and all the water throughout Egypt turned to blood. This marked the first of the ten

---

[135]See https://en.wikipedia.org/wiki/List_of_prizes_for_evidence_of_the_paranormal

[136]Walter Brown Gibson (September 12, 1897 – December 6, 1985) was an American author and professional magician, best known for creating and developing the pulp magazine character *The Shadow*. Writing under the pen name Maxwell Grant, Gibson authored more than three hundred Shadow stories, making him one of the most prolific writers in the genre.

plagues unleashed upon the land—a divine response to Pharaoh's hardened heart and his refusal to release the Israelites.

Remarkably, the Egyptian magicians managed to replicate this phenomenon by turning other water sources red. This imitation emboldened Pharaoh. If his own sorcerers could match the signs performed by Moses and Aaron, then he had no reason to fear their God. He saw their acts not as divine wonders, but as mere tricks—tricks his own men could eventually master.

The second plague came in the form of an overwhelming infestation of frogs. Once again, the Egyptian magicians succeeded in reproducing the effect, conjuring their own swarm of batrachians. Up to that point, Pharaoh had encountered nothing to convince him that these plagues were beyond human imitation or rooted in true divine power. If his court magicians could keep up, he felt secure in his skepticism.

But that confidence would soon wane. As the plagues continued, the magicians found themselves unable to replicate them. Their powers—or illusions—fell short. Still, the initial successes were enough to cast doubt on the legitimacy of Moses and Aaron, allowing Pharaoh to dismiss them as mere performers of a more advanced magic.

It was not until the final plague—the death of all the firstborn—that Pharaoh's resolve was shattered. This last blow was devastating, unanswerable, and absolute. No spell or incantation could undo what had been done. There was no illusion, no trick, no magic that could bring the dead back. It was the end of all ends. And in the silence that followed, Pharaoh finally relented, conceding to the will of a God he could no longer deny.

Lawrence Alma-Tadema, a Dutch painter of the Victorian era, was known for his meticulously detailed and opulent neoclassical works. Trained in Belgium and residing in England from 1870, Alma-Tadema gained renown for his vivid portrayals of the ancient world. Among his most celebrated paintings is *The Death of the Pharaoh's First Son*[137], currently housed in the Rijksmuseum in Amsterdam.

---

[137]See https://www.rijksmuseum.nl/en/search?q=SK-A-2664

In this powerful composition, Alma-Tadema depicts the Pharaoh of Egypt holding the lifeless body of his firstborn son on his lap. The queen clings to her child in a posture of raw despair, her face etched with anguish. Around them, servants mourn in silence while dancers enact the ritual dance of death. The dim, flickering candlelight casts long shadows across the scene, amplifying its tragedy and emotional intensity.

At the heart of the canvas stands Pharaoh himself—commanding, robbed in the regalia of his rank, adorned with the symbols of his power. Yet, despite his majestic bearing, the presence of his son's corpse in his arms betrays a stark vulnerability. The boy's cyanotic skin, with its bluish lips and fingernails, contrasts cruelly with the gleam of the golden scarab amulet he wears—a sacred talisman meant to protect, now rendered powerless. Its failure is as symbolic as it is visible.

To the left, partially obscured in the shadows, stand Moses and Aaron—the emissaries of the God of Israel. Their grim expressions and silent presence remind Pharaoh that their prophecy has happened. They await the words that will soon be spoken: "Hebrews, you may leave Egypt."

To the right, another figure draws the viewer's attention: a court physician collapsed on the ground in defeat. Before him lies an array of balms and ointments—his entire arsenal of healing, now useless. His posture, bent and broken, mirrors his emotional state: impotent, overwhelmed by a death he cannot comprehend, much less cure.

Alma-Tadema's work masterfully captures a universal truth: death humbles even the greatest of rulers. It reduces wealth, knowledge, and magic alike to irrelevance. The finality of death is what humanity has always feared most—for it robs us of tomorrow and buries all our hopes in the grave.

And yet, the painting—so steeped in resignation—leaves one lingering question. Is death truly the end? Is the dead forever gone? Or can the Master of Life—who issued the judgment—also reverse its decree? Might there be an exception to this most unyielding of laws?

ARGUMENT: JESUS CHRIST IS RISEN INDEED!

Dad, Mom, and their two children—one ten years old, the other seven—were full of excitement as they settled into their new home. After unpacking and organizing most of their belongings, they decided it was time to refresh the interior with new colors. They had already discussed their vision, so without delay, they purchased the paint and got to work.

They began with the main room, determined to finish it that very day, no matter how late it got. Immersed in the task, time slipped away unnoticed. Eventually, the father glanced at his watch and was startled to see that it was midnight—and the kids were still up.

Just then, the younger child walked into the room. The father asked him where his brother was. "Watching television," the boy replied. Without hesitation, the father instructed, "Tell him to turn off the TV immediately and both of you go to bed."

Obediently, the younger son went to relay the message. Approaching his older brother, he said, "Dad says to turn off the television right now and go to sleep."

At that moment, the older boy found himself caught in a dilemma. He considered the possibility that his younger brother might be making it up—as a clever trick to get him to shut off the television, just so he could sneak back and watch his favorite program once the coast was clear.

But what if the message was real? What if his brother was truly acting as a messenger of their father's authority? Ignoring the command could get him in trouble.

How could he be sure the message was genuine? How could he know he could trust the messenger?

Common sense often tells us that the dilemma between obeying or disobeying a command should be easily resolved by the authority of the one who issues it. If the source is credible and legitimate, the expectation is that the command will be followed. And yet, from the very beginning of human history, we encounter examples of disobedience—even in the face of unquestionable authority.

In the Garden of Eden, God gave a clear and solemn command to our first parents, Adam, and Eve: they were not to eat from the tree of the knowledge of good and evil, for if they did, they would surely die. A simple, unequivocal directive from the Creator Himself. Yet the serpent entered the scene and contradicted God's word, assuring Eve that they would not die and, in fact, would become like gods, knowing good and evil.

At that moment, a choice had to be made. Whom should they believe—God, the omniscient Creator, or the serpent, a creature with an enticing but contradictory message? On the surface, the right choice seems obvious. And yet, we know what Adam and Eve chose. We also know the consequences of that fateful decision—an act of disobedience that echoed through history.

Centuries later, another moral and spiritual crisis unfolded—this time in the shadowed garden of Gethsemane and in the courts of human judgment. Jesus, arrested and brought before the Jewish Sanhedrin on charges of blasphemy for declaring Himself the Son of God, was condemned and sent to the Roman governor Pontius Pilate to authorize His execution.

Pilate questioned Jesus in the Praetorium, as recounted in the Gospel of John (18:33–38). He asked, "Are you the King of the Jews?" Jesus responded not with a direct answer, but with a probing question of His own: "Are you asking this on your own, or have others told you about me?"

Pilate, unsettled, replied, "Am I a Jew? Your own people and chief priests have handed you over to me. What is it you have done?"

Then Jesus offered a profound answer: "My kingdom is not of this world. If it were, my servants would fight to prevent my arrest by the Jewish leaders. But now my kingdom is from another place." (John 18:36).

The charge of blasphemy meant little to the Roman governor, Pontius Pilate. In the context of pagan Rome—where a pantheon of gods flourished and divine claims were common—the idea of someone proclaiming himself to be the Son of God was not particularly alarming. To Roman ears, such accusations often sounded trivial, even absurd. But when Jesus spoke of a "kingdom," Pilate's attitude shifted. This was no longer a matter of religious semantics—it now touched the realm of politics and imperial authority, a matter that very much fell under Roman jurisdiction.

So, Pilate pressed further: "Then you are a king?" Jesus not only affirmed the title but also unveiled the purpose of His entire life: "I was born and came into the world to testify to the truth" (John 18:37).

In this exchange, we encounter a man who boldly claims to be the Son of God—a man who fulfills the ancient prophecies concerning the Messiah. But how can we verify such authority? How can we know that He truly was God's messenger?

As God declared to Moses in Deuteronomy 18:15–22, a prophet is either true or false—there is no middle ground. If every prophecy that pointed to the coming of the Messiah was fulfilled in the person of Jesus, then the prophetic foundation validating His identity holds firm. And if those prophets were proven true by the fulfillment of their words, what basis remains to reject the one to whom they pointed?

Still, Jesus' miracles—extraordinary though they were—were not presented by Him as the definitive proof of His divine nature. When the Pharisees and Sadducees demanded a sign, He did not cite the raising of Lazarus, or the feeding of thousands with five loaves and two fish. He did not mention giving sight to the blind or making the paralyzed walk. He offered none of these as proof.

Instead, He pointed to one singular event: His resurrection. "No sign will be given," He said, "except the sign of Jonah"—a reference to His rising from the grave after three days. That, He declared, would be the ultimate validation of His identity. Unlike any prophet before Him, Jesus claimed that He would conquer death itself.

And so, the resurrection stands as the cornerstone of Christianity. If Jesus rose from the dead, then every word He spoke is vindicated, every promise confirmed. But if He did not—if the resurrection is a lie, a fabrication, or a delusion—then the entire edifice of Christian faith collapses.

This is the hinge upon which everything turns. As the Apostle Paul wrote, "If Christ has not been raised, our preaching is useless and so is your faith" (1 Corinthians 15:14).

More than two thousand years have passed, and no evidence has emerged to decisively disprove the resurrection. On the contrary, growing volumes of historical, textual, and experiential testimony continue to support it. Far from being discredited, the resurrection of Jesus remains the most examined, defended, and enduring claim in human history—and the heart of Christian belief.

The meaning of Jesus' resurrection lies in the realm of theology. But the disappearance of His body—that is a matter for historical investigation. To classify any event as historical, it must satisfy two essential conditions: it must have occurred at a specific time and in a specific place. Without these anchors in space and time, an event remains speculative, outside the domain of verifiable history.

The resurrection of Jesus meets both criteria. His burial took place in a tomb hewn into the rock of a hillside near Jerusalem. The timing is well established: it occurred during the prefecture of Pontius Pilate, who governed the Roman province of Judea between AD 26 and 36. There is a definite where and a definite when.

The historical reliability is further reinforced by the presence of verifiable individuals involved in the burial and trial of Jesus. Joseph of Arimathea, a wealthy and influential member of the Sanhedrin, is attested not only in the Gospels but in extra-biblical sources as a real figure in Jewish leadership. He offered his personal tomb for Jesus' burial. Nicodemus, another prominent Sanhedrin member, assisted in the burial, bringing with him an extraordinary amount of embalming spices—about one hundred pounds of myrrh and aloes. He, too, is referenced in several apocryphal writings and was later associated with a burial site next to the tomb of Saint Stephen, discovered in AD 415.

Joseph ben Caiaphas, the high priest who presided over Jesus' trial, is also a historically confirmed individual. His ossuary, or bone box, was discovered in Jerusalem and is currently displayed at the Israel Museum. His house's remains have been unearthed and can still be visited in the city. Bronze coins minted in Galilee between AD 26 and 36 further confirm the timeline of Pilate's governance, synchronizing with the Biblical record.

All of this converges to a striking point: every major figure involved in the Passion and crucifixion of Jesus is historically attested, not only by Christian sources, but through secular records and archaeological findings. These are not mythical characters. They lived, governed, acted, and left traces in the historical and archaeological record.

We know where the bones of Abraham, Mohammed, Buddha, Confucius, Lao-Tzu, and Zoroaster are. But where are those of Jesus? The nature of the resurrected Jesus' body may remain a mystery, but the fact of his disappearance is a matter to be decided by historical evidence, such as that which I will present later.

All the evidence found in the New Testament and in early Church writings clearly demonstrates that the proclamation of the Gospel was not simply, "Follow the Master's teachings and be good," but emphatically, "Jesus Christ has risen from the dead." The resurrection cannot be removed from Christian doctrine without radically altering its character and destroying its true essence.

As mentioned in the previous chapter, God gave the Israelites a clear standard for identifying a true prophet (Deuteronomy 18:21–22): if what the prophet proclaims in the name of the Lord comes to pass, he is a true prophet; otherwise, he is not. Jesus, like the prophets of antique, foretold many things: his betrayal, passion, death, resurrection, the persecution of his followers, and even the destruction of Jerusalem.

> From that time onward, Jesus began to make it clear to his disciples that He had to go to Jerusalem and endure much suffering at the hands of the elders, the chief priests, and the scribes. He would be put to death, and on the third day He would be raised up. (Matthew 16:21)

As Jesus was going up to Jerusalem, He took the twelve disciples aside by themselves and said to them on the way, 'We are going up to Jerusalem, and the Son of Man will be handed over to the chief priests and the scribes. They will condemn him to death and hand him over to the Gentiles to be mocked, scourged, and crucified, and on the third day He will be raised up.' (Matthew 20:17–19)

After saying this, Jesus was deeply troubled and He declared, 'Amen, amen, I say to you, one of you will betray me.' [...] Jesus answered, 'It is the one to whom I will give a piece of bread after I have dipped it.' So, when He had dipped the piece of bread, He gave it to Judas, son of Simon Iscariot. (John 13:21, 26)

Jesus said to him, 'Amen, I say to you, this very night, before the cock crows, you will deny me three times.' (Matthew 26:34)
Be on your guard, for they will hand you over to the courts and you will be beaten in the synagogues. You will be brought before governors and kings because of me, as a testimony to them. The gospel must first be proclaimed to all nations. (Mark 13:9–10)

As Jesus was leaving the temple, one of his disciples said to him, 'Teacher, look at the tremendous size of these stones and these magnificent buildings!' Then Jesus said to him, 'Do you see these enormous buildings? Not a single stone will be left upon another; everyone will be thrown down.' (Mark 13:1–2)

Jesus, being thoroughly familiar with the Scriptures, would have known the words spoken by God through the prophet Ezekiel concerning false prophets:

Therefore, thus says the Lord God: Because you have spoken falsehood and had lying visions, I am against you, says the Lord God. My hand will be against the prophets who see false visions and utter lying divinations. [...] Because they lead my people astray, saying, 'Peace,' when there is no peace, and when a flimsy wall is built, they cover it with whitewash. [...] I will tear down the wall that you covered with whitewash and level it to the ground so that its foundation is laid bare. (Ezekiel 13:6–14)

All the prophecies made by Jesus were fulfilled, including the dramatic and unlikely destruction of the Temple in Jerusalem. At the time, few would have believed such devastation was possible. The Temple was a massive structure—approximately five hundred meters long and three hundred meters wide—built with enormous stone blocks weighing several tons. It stood as a formidable fortress and a symbol of religious and national identity for the Jewish people.

Yet, in AD 66, the Jewish population rose in rebellion against the Roman Empire. Four years later, in AD 70, following a grueling siege of over five months, Roman legions under the command of Titus—on behalf of his father, Emperor Vespasian—destroyed much of Jerusalem. The Second Temple, the heart of Jewish worship, was reduced to ruins. The Arch of Titus, still standing in Rome today, commemorates that victory and depicts Roman soldiers carrying away sacred Temple artifacts, including the Menorah.

Through His resurrection, Jesus definitively demonstrated that He was neither deluded nor deceitful in claiming to be the Son of God. On the contrary, He was the true Messenger of the Father, come to fulfill and give new depth to what the prophets had foretold. In Him, all the Scriptures were confirmed—He spoke them, explained them, and fulfilled them.

No mere man preaching "truth" would so constantly appeal to the Scriptures unless they were themselves true. In fact, Jesus seemed intensely devoted to them. He referred to them frequently, drawing wisdom, instruction, and authority from their words. His teachings were saturated with Scripture, and He seized every opportunity to reveal their divine wisdom and prophetic fulfillment.

When the Lord spent forty days in the wilderness and was tempted by the devil, He responded each time by quoting Scripture: "He answered, 'It is written, One does not live by bread alone but by every word that comes forth from the mouth of God.'" [138] (Matthew 4:4). "Jesus said to him, 'Again it is written, you shall not put the Lord your God to the test.'" [139] (Matthew 4:7). "Then Jesus said to him, 'Away with you, Satan! For it is written, you shall worship the Lord your God and serve him alone.'" [140] (Matthew 4:10)

---

[138]Deuteronomy 8:3.

[139]Deuteronomy 6:16-18.

[140]Deuteronomy 6:13.

For Jesus, every question or challenge found its answer in the Scriptures. When questioned about working on the Sabbath, He replied: "I ask you: Is it lawful on the Sabbath to do good or to do harm, to save life or to destroy it?" (Luke 6:9). When asked what one must do to inherit eternal life, He pointed directly to Scripture: "What is written in the Law? What do you read there?" [...] He answered, 'You shall love the Lord your God with all your heart, with all your soul, with all your strength, and with all your mind, and your neighbor as yourself.' Jesus said to him, 'You have answered correctly. Do this, and you will live.'" (Luke 10:26–28). And when asked about the greatest commandment in the Law, Jesus quoted Deuteronomy: "'Teacher, which commandment in the Law is the greatest?' Jesus said to him, 'You shall love the Lord your God with all your heart and with all your soul and with all your mind.'" (Matthew 22:36–37).

Jesus consistently affirmed His divine authority by referring to the Scriptures as the ultimate source of truth and guidance. On numerous occasions, He rebuked those who failed to understand or even read them. When He drove the merchants out of the Temple, He declared: "It is written: My house shall be called a house of prayer, but you are making it a den of thieves."[141] (Matthew 21:13). At the conclusion of the parable of the wicked tenants, He challenged the religious leaders: "Have you never read in the Scriptures: The stone that the builders rejected has become the cornerstone. By the Lord this has been done, and it is wonderful in our eyes?"[142] (Matthew 21:42). Jesus frequently referenced the Scriptures in His teachings, underscoring their divine authority. He once admonished His listeners, saying: "You search the Scriptures because you think that in them you have eternal life. Yet it is they that testify about me, but you are not willing to come to me to have life" (John 5:39–40). Even His critics were struck by His profound scriptural knowledge: "The Jews were astonished and said, 'How is it that this man has such learning when he has never studied?'" (John 7:15). When inviting people to place their faith in Him, Jesus once again grounded His appeal in Scripture: "Let anyone who is thirsty come to me. Let the

---

[141]Isaiah 56:7.

[142]Psalms 118:22

one who believes in me drink. As the Scripture says, 'Out of the believer's heart shall flow rivers of living water.'" (John 7:37–38).

One must ask: How could Jesus have used the Scriptures so effectively—to correct error, rebuke wrongdoers, guide the lost, educate the ignorant, and resist temptation—if they were not truly the Word of God?

Never once did Jesus diminish the authority of the Scriptures. On the contrary, He affirmed their enduring validity and elevated their role in salvation history. In His own words:

> Do not think that I have come to abolish the Law or the Prophets. I have come not to abolish but to fulfill. For amen, I say to you, until heaven and earth pass away, not the smallest letter or the smallest part of a letter will pass from the Law until all is accomplished. (Matthew 5:17–18)

If it is demonstrated that Jesus truly died and rose again—that He did not remain in the tomb but returned to life—then it is also demonstrated that He is indeed the Son of God, the One whom the Father sent to reveal His will and communicate His voice to humanity. The resurrection confirms beyond doubt that everything Jesus said is true, that His words are trustworthy, and that His teachings carry divine authority.

By rising from the dead, Jesus placed the seal of truth upon the Scriptures. His resurrection validates not only His identity but also affirms the Bible as the authentic Word of God. If Christ trusted, quoted, fulfilled, and affirmed the Scriptures—and if His resurrection proves His divine nature—then we, too, can have complete confidence in the way God has chosen to speak to us through the Sacred Texts.

FIRST THESIS: DEATH IN THE TIME OF JESUS

The way in which death is mourned in the modern Western world contrasts sharply with the customs and attitudes of the ancient Near East. To understand the significance of Jesus' burial—and the cultural weight surrounding it—it is essential to first explore the funerary practices and expressions of grief in the region during His time.

In first-century Jewish culture, the death of a loved one was met with a profound and emotionally charged response. Mourning began immediately and was often expressed through intense, even visceral lamentation. Ancient sources describe this initial reaction as a high-pitched, ear-piercing wail—a public outcry of sorrow and despair. This dramatic expression of grief had deep roots in Jewish memory, recalling the night of the first Passover in Egypt: "Pharaoh arose during the night, he and all his officials and all the Egyptians, and there was a loud cry of grief in Egypt, for there was not a single house in which someone was not dead." (Exodus 12:30)

This collective mourning became a ritualized aspect of Jewish bereavement. Family and friends would continue lamenting from the moment the first wail was heard until the deceased was buried, typically within 24 hours. These public laments were not merely emotional outbursts but a communal act of honoring the dead and expressing solidarity in grief.

This custom is vividly illustrated in the Gospel account of Jairus' daughter. When Jesus arrived at the synagogue leader's home, He encountered the characteristic scene of a Jewish household in mourning: "When they arrived at the house of the synagogue official, Jesus noticed a commotion, with people weeping and wailing loudly." (Mark 5:38)

Such scenes were common in Jewish funerary settings and would have surrounded Jesus' own death and burial. Recognizing these deeply rooted traditions allows us to appreciate not only the emotional atmosphere at the time of Jesus' death, but also the theological and cultural weight His burial carried.

Among the mourning customs of ancient Israel, one notable practice was the hiring of professional mourners, typically women, whose role was to publicly lament and express grief on behalf of the bereaved. These women would wail, cry out, and sometimes sing dirges, heightening the emotional atmosphere of funerals and times of communal sorrow. The prophet Jeremiah directly refers to this custom: "Call for the mourning women to come; send for the most skillful of them. Let them hasten and raise a lament for us so that our eyes may overflow with tears and our eyelids run with water." (Jeremiah 9:17–18).

This vivid, poetic portrayal reflects how ingrained these lamenting rituals were in the fabric of Israelite mourning culture.

Another expressive element of grief was the use of sackcloth (cilicio), a coarse, dark fabric typically made from camel or goat hair. Sackcloth was used to make rough garments—either worn alone or placed over existing clothing—as a visible sign of sorrow and penitence. This practice is the origin of the black mourning attire common in many cultures today.

For example, when Abner, commander of Saul's army, was killed, King David ordered the people to mourn using traditional signs of grief: "Then David said to Joab and all the people who were with him, 'Tear your garments, put on sackcloth, and mourn over Abner.' And King David followed the bier." (2 Samuel 3:31) The tearing of garments was another powerful symbol of intense sorrow or outrage. It was a public display of internal anguish, used in both personal and communal crises. We see this custom vividly in the Passion narrative, when Caiaphas, the high priest, reacts to Jesus' declaration of His divine identity: "Then the high priest tore his garments and said, 'He has blasphemed! What further need do we have of witnesses? You have now heard the blasphemy for yourselves.'" (Matthew 26:65)

In ancient Jewish tradition, it was customary to bury the dead quickly, typically on the same day of death. This urgency was driven by two primary reasons. First, the hot and arid climate of the Middle East led to rapid decomposition of corpses. Second, and perhaps more importantly from a cultural standpoint, delaying burial was considered a dishonor to both the deceased and the surviving family members.

This practice is evident in several biblical accounts. The Gospels and the Book of Acts record at least three instances where burial occurred on the very day of death:

- Jesus was buried the same day He died: "Joseph took the body, wrapped it in a clean linen shroud, and laid it in his own new tomb that he had hewn in the rock." (Matthew 27:59–60)
- Ananias, after lying to the apostles, died suddenly and was buried at once: "When Ananias heard these words, he fell down and died. [...] The young men came forward, wrapped up his body, and carried him out for burial." (Acts 5:5–6)
- Stephen, the first Christian martyr, was also buried promptly after being stoned: "Devout men buried Stephen and made loud lamentation over him." (Acts 8:2)

The custom of same-day burial also appears much earlier in salvation history. When Rachel, Jacob's beloved wife, died during a journey, she was not brought back to be buried in the family tomb but was interred immediately: "Thus Rachel died, and she was buried on the road to Ephrath (now Bethlehem)." (Genesis 35:19).

Moreover, Jewish Law explicitly required the burial of executed criminals before nightfall, underscoring the dignity owed even to the condemned: "If a man is guilty of a capital offense and is put to death, and you hang him on a tree, his body must not remain there overnight. Be sure to bury him that same day." (Deuteronomy 21:22–23). This commandment was fulfilled in the case of Jesus, who—though crucified as a criminal—was buried before sunset in accordance with the Law.

Ancient Jewish belief also held that the spirit of a deceased person lingered near the body for three days, listening to the mourners and remaining somehow "close." After this period, it was thought that the spirit departed entirely and hope for restoration was lost. This cultural perspective appears in the narrative of Lazarus, when Martha, his sister, expresses her despair to Jesus: "Martha, the sister of the dead man, said to him, 'Lord, by now there will be a stench, for he has been dead for four days.'" (John 11:39)

To this day, certain ancient burial customs remain in practice in parts of the Middle East, particularly in Syria. Among these traditions is the wrapping of the dead: the face is first covered with a scarf, and then the head, hands, and feet are wrapped in strips of linen cloth. In some cases—especially if the deceased was a person of status—the linen used may have originally been designated for wrapping sacred scrolls of the Law. Once wrapped, the body is carried to the grave and buried.

This practice sheds light on the resurrection of Lazarus, whose burial followed these customs. When Jesus called him from the tomb, Lazarus appeared still wrapped in the traditional funeral cloths: "The man who had died came out, his hands and feet bound with strips of cloth, and his face wrapped in a cloth. Jesus said to them, 'Unbind him, and let him go.'" (John 11:44)

The use of spices and aromatic substances during burial was optional and expensive, typically reserved for the wealthy. These materials helped mask the odor of decomposition and served as a sign of honor. Initially, myrrh and aloes were used; in later periods, other elements such as hyssop, perfumed oils, and rose water became more common.

A complete linen wrapping—such as the one used for Jesus—was not universally practiced but signified dignity and reverence. According to the Gospels, Joseph of Arimathea, a wealthy disciple, provided his own tomb and burial materials for Jesus, including an abundance of spices (John 19:39–40).

Tombs of that era were typically carved into rock and included a bench-like projection where the body was laid during decomposition. Once the flesh had decayed, the bones were collected and placed into an ossuary—a small container made of stone or clay. These ossuaries required little space and were often kept in family tombs, allowing the same burial site to be reused for generations.

Because of this repeated use, tombs were built with openable entrances. A large stone was rolled into place to cover the opening and protect the tomb, like the one that sealed Jesus' burial place—a tomb belonging to Joseph of Arimathea.

It was also customary to whitewash the exterior of tombs during springtime, especially before the Passover. This made the tombs more visible and served to prevent ritual impurity, since accidentally touching a grave rendered a person unclean according to Jewish law (Numbers 19:16). This background illuminates Jesus' powerful rebuke of the Pharisees: "Woe to you, scribes and Pharisees, hypocrites! You are like whitewashed tombs which appear beautiful on the outside but inside are full of the bones of the dead and all kinds of filth." (Matthew 23:27)

Had Jesus been buried like any ordinary stranger or pilgrim who died in Jerusalem, His body would have been placed in a simple grave in the ground, sealed permanently and never opened again. In such a case, the physical evidence of His resurrection—such as the stone rolled away and the linen burial cloths left behind—would not have been so striking or verifiable. These tangible signs gave early witnesses compelling confirmation that something extraordinary had occurred.

The open tomb and the remaining burial cloths served as silent but powerful testimonies to His resurrection, as foretold in Scripture. I will return to this point in greater depth later.

Historically, such ground burials were customary for servants, strangers, or the poor. For instance, Deborah, Rebekah's nurse, was buried beneath an oak tree: "Deborah, Rebekah's nurse, died and was buried under the oak below Bethel." (Genesis 35:8). Many times, natural caves serve as family tombs. The Cave of Machpelah, for example, became the burial site for Abraham, Sarah, Isaac, Rebekah, Leah, and Jacob: "There they buried Abraham and his wife Sarah; there they buried Isaac and his wife Rebekah, and there I buried Leah." (Genesis 49:31). Only prophets and kings were typically buried within city limits, which was seen as an honor. Samuel was buried at his home in Ramah: "Then Samuel died, and all Israel assembled and mourned for him. They buried him in his home at Ramah." (1 Samuel 25:1). Similarly, David was buried in the city of Jerusalem: "David rested with his ancestors and was buried in the City of David." (1 Kings 2:10).

In contrast, the poor were buried in a communal cemetery located outside the city walls, as noted in the reforms of King Josiah: "He removed the bones from their graves and burned them on the altar to desecrate it, in accordance with the word of the Lord. Then he returned to Jerusalem." (2 Kings 23:6).

Jesus, however, received a burial befitting a man of great honor and wealth. The use of a linen shroud, the application of about a hundred Roman pounds (roughly 33 kilograms) of myrrh and aloes, and the placement of His body in a new tomb carved from rock—all indicate a burial of the highest dignity: "They took the body of Jesus and wrapped it with the spices in linen cloths, according to the burial custom of the Jews." (John 19:40)

The tomb was donated by Joseph of Arimathea, a wealthy and respected member of the Sanhedrin, while Nicodemus provided the costly spices. These were not items that ordinary people kept on hand—especially at the onset of the Sabbath, when purchasing such materials would have been impossible.

What's striking is that all burial rituals customary for the recent deceased were carefully observed in Jesus' case. Yet no one involved in the burial anticipated that He might rise from the dead, despite the prophetic affirmations in the Psalms and in Jesus' own words. The women and disciples who performed the burial rites believed they were preparing a body destined for long-term decay, not resurrection.

This makes the resurrection accounts even more credible. Why go through the costly, elaborate process of burial if the expectation were that Jesus would rise in three days? Why use expensive linen and perfumes if His return was imminent? The fact that they did shows they were not expecting an empty tomb. Only Mary, His mother, may have held that hope quietly in her heart.

Although the Gospels record no words from Mary during the burial, one can imagine that she recalled His repeated prophecies about His passion and resurrection. She shared them with those preparing the body, though they—like most others—did not believe her. This may explain why Mary did not participate in the burial rituals or accompany the other women on the morning of the resurrection. She may have been the only one who truly believed He would rise again.

SECOND THESIS: MULTIPLE WITNESSES

William Jordan, a retired sergeant of the Los Angeles Police Department, was one of the officers assigned to investigate the assassination of Senator Robert Francis Kennedy, commonly known as Bobby Kennedy. The tragic event occurred in the early morning hours of June 5, 1968, just after Kennedy had delivered his victory speech following his win in the California Democratic primary.

In an interview aired on the *History Channel*, Sergeant Jordan recalled that one of the greatest challenges of the investigation was the sheer number of witnesses present. There were hundreds of eyewitnesses, and every one of them offered a different version of what had occurred. All of them had heard the shots that fatally wounded the Senator from New York, and many even claimed to have seen the assailant. Yet each account included a multitude of conflicting details, many of which turned out to be irrelevant.

Still, because of the high profile of the victim, every testimony had to be recorded and considered. The investigative team was required to treat each account with seriousness, even when contradictions emerged, as each person's perspective was shaped by their vantage point and emotional state during the chaos.

Importantly, the existence of differing versions of the same event does not mean that the witnesses were lying. On the contrary, such variation is normal and expected in eyewitness testimony. In fact, a judge would become suspicious if every witness gave the same account, down to the smallest details. That kind of uniformity would suggest collusion, not honesty.

Human beings perceive and remember events through unique lenses, especially when the moment carries a high emotional charge. It is precisely this diversity of accounts that gives credibility to the overall narrative. If all the witnesses had told an identical story, it would more likely have indicated that they had coordinated their testimonies, thereby intentionally misleading the investigation.

The resurrection of Jesus is narrated by all four evangelists— Matthew, Mark, Luke, and John. Each Gospel provides unique details that the others do not include, allowing us to construct a fuller and more vivid picture of this monumental event. Given the nature of eyewitness testimony and differing perspectives, it is neither expected nor necessary for the accounts to be identical. As previously noted, such variation is characteristic of truthful narratives rather than coordinated fabrications.

Despite the differences in detail, the core facts remain consistent across all four Gospels: Jesus' tomb was found empty, He had risen, and He appeared to His followers. This convergence of essential truths across independent sources offers strong grounds for confidence in the historical credibility and theological authenticity of their testimony.

The resurrection narratives begin at dawn on the first day of the week. A group of women—among them Mary Magdalene—goes to the tomb with the intention of completing the anointing rituals that could not be finished before the Sabbath. However, upon arriving, they found the stone rolled away and the tomb empty.

At that moment, angels appear to the women and announce that Jesus has risen from the dead. The women are overcome with a mixture of fear and joy, astonished by what they have witnessed and heard. The chronology of the events that follow becomes difficult to pinpoint with precision—an expected feature of multiple independent eyewitness reports. However, a general sequence can be discerned.

Mary Magdalene, running ahead of the other women, goes to inform Peter and John of what she believes is the disappearance of the Master's body. The two apostles, along with Mary, hurry to the tomb. John arrives first, but it is Peter who enters. They found only the linen clothes, and the tomb empty, just as Mary had said. Deeply moved and perplexed, they returned home, still unsure of what to make of it all.

Left behind, Mary Magdalene remains near the tomb, weeping. It is then that Jesus appears to her, though at first, she does not recognize Him. Upon realizing it is the Risen Lord, she becomes the first recorded witness of the resurrection. Shortly after this, Jesus also appears to the other women, who are on their way to inform the disciples of what they had seen and heard at the tomb.

And yet, in a striking detail shared across the Gospels, the disciples do not believe them. Their words seemed, as Luke recounts, "like nonsense" to the men (cf. Luke 24:11). This skepticism only underscores the unexpected nature of the resurrection, and paradoxically, it lends credibility to the accounts. If the evangelists were inventing a story, they would be unlikely to portray the apostles—future leaders of the Church—as doubtful and slow to believe.

There is no detailed biblical record of the personal encounter between Jesus and Peter on the day of the resurrection. The only explicit reference comes from Paul, who, in listing the appearances of the Risen Lord, briefly states: "And that He appeared to Cephas, and then to the Twelve." (1 Corinthians 15:5).

This sparse mention leaves the nature of that meeting to our imagination, yet its inclusion in Paul's list—and its quiet acknowledgment by the other disciples—indicates that it held deep significance for Peter.

The Gospel of Luke also alludes to this appearance indirectly. When Cleopas and his companion return from their encounter with the Risen Christ on the road to Emmaus, the other disciples greet them with these words: "The Lord has truly been raised and has appeared to Simon!" (Luke 24:34).

Following this, Jesus appears to the rest of the apostles, although Thomas is notably absent during this first encounter (John 20:24). The four Gospels, when considered together can be arranged into a probable chronological sequence, as follows:

- The stone is rolled away from the tomb (Matthew 28:2–4)
- Several women arrive at the tomb (Mark 16:1–4; Matthew 28:1; Luke 24:1–3; John 20:1)
- Angels announce the resurrection to the women (Mark 16:5–7; Matthew 28:5–7)
- The angels remind the women of Jesus' prophecy (Luke 24:4–8)
- The women flee the tomb, trembling and astonished (Mark 16:8)
- Mary Magdalene informs Peter and John (John 20:2)

- Peter and John run to the tomb and examine it (John 20:3–10; Luke 24:12)
- Jesus appears to Mary Magdalene (Mark 16:9; John 20:11–17)
- Jesus appears to the other women as they return from the tomb (Matthew 28:8–10)
- The guards report the event to the chief priests, who devise a cover-up (Matthew 28:11–15)
- The women tell the disciples what they witnessed (Luke 24:9–11; Mark 16:10–11; John 20:18)
- Jesus appears to Peter (Cephas) (Luke 24:34; 1 Corinthians 15:5)
- Jesus appears to Cleopas and another disciple on the road to Emmaus (Luke 24:13–27; Mark 16:12)
- Jesus reveals Himself to them in the breaking of the bread (Luke 24:28–32)
- Cleopas and his companion return to Jerusalem and recount their encounter (Luke 24:33–35; cf. John 20:19; Mark 16:13)
- Jesus appears to the gathered disciples (without Thomas) (Luke 24:36–44; John 20:19–20; Mark 16:14)

While it is entirely reasonable to expect certain differences among the resurrection narratives, a casual or unfamiliar reader might interpret these differences as contradictions, even suspecting one or more of the evangelists of error or fabrication. This concern, however, is misplaced. With careful reading and logical reasoning, these apparent discrepancies can be reconciled in ways that enhance the credibility of the accounts rather than diminish it.

Let us consider one example: the number of angels present at the tomb. In Matthew 28:2–7, the evangelist refers to a single angel, while Luke 24:4–7 mentions two angels. This seems inconsistent. But a closer look reveals that Matthew does not say there was only one angel; he simply chooses to focus on one—possibly because only one of the angels spoke, or because he wished to emphasize that particular figure's role. Luke, on the other hand, offers a fuller description by noting that two men in dazzling garments appeared.

There is no contradiction here—only a difference in emphasis. If there were two angels, it is perfectly reasonable for one narrator to highlight only the speaking angel, especially in a moment of high theological symbolism and dramatic intensity. As with modern eyewitnesses, each evangelist recorded what stood out most to him or to the community he wrote for, without denying the fuller context.

Another commonly cited difference involves the earthquake described in Matthew's account: "Suddenly there was a great earthquake, for an angel of the Lord descended from heaven and coming to the tomb, rolled back the stone and sat on it." (Matthew 28:2,).

None of the other evangelists—Mark, Luke, or John—mention this earthquake. Does that mean Matthew invented it? Not at all. The absence of a detail in another account does not imply fabrication. It simply means that the other writers chose to leave it out, likely because their focus was directed toward other theological or narrative elements.

In 2012, Spanish journalist Pepe Rodríguez, a well-known critic of the Catholic Church, published a provocative book titled *Fundamental Lies of the Catholic Church*. The book became a commercial success, resonating with many readers who were skeptical of Christian belief, especially the resurrection of Jesus. In his work, Rodríguez systematically challenges the Gospel accounts of the resurrection, pointing to what he calls contradictions among the four narratives and using them to argue that the resurrection is a fabricated story. Based on these claims, he concludes that the event never happened and that the Gospel testimonies cannot be trusted.

Unfortunately, many readers accepted his conclusions without a deeper examination of the nature of eyewitness testimony, the purpose of the Gospels, or the contextual coherence of the resurrection narratives. His critique relies on the assumption that any difference between Gospel accounts is a contradiction—an assumption that is both historically naïve and logically flawed.

However, as we have seen in the previous examples, the so-called discrepancies—such as the number of angels at the tomb or the mention of the earthquake—are not contradictions at all. Rather, there are variations in emphasis and detail, which are not only natural but expected in authentic accounts of profound, emotionally charged events.

Eyewitnesses often highlight different aspects of the same event based on their perspective, memory, and the message they seek to convey.

Another compelling reason to trust in the credibility of the evangelists' accounts lies in a detail they could never have fabricated convincingly: the mysterious transformation of Jesus' resurrected body.

Let me explain. Prior to His Passion and resurrection, the Gospels record three resurrections performed by Jesus in the presence of His disciples:

- The resurrection of Lazarus, His close friend (John 11:1–44)
- The raising of the widow of Nain's only son (Luke 7:11–17)
- The resurrection of Jairus' daughter, the child of a synagogue leader (Matthew 9:18–25; Mark 5:21–43)

In each of these cases, the person who had died returned exactly as they had been before death. Their friends and family immediately recognized them, and life resumed as before. These were miraculous restorations, but they were resurrections to mortal life—the individuals would live again, but still age and eventually die once more. The disciples, having witnessed these events, naturally formed a conceptual framework: resurrection meant the return of life to a lifeless body, with no change in identity, form, or appearance.

If the evangelists were inventing or embellishing the resurrection of Jesus, it would have made sense for them to model it on what they already knew: the familiar pattern of recognizable resurrections. But that is not what they record. Instead, we are told again and again that those who encountered the risen Jesus did not immediately recognize Him—despite the clear evidence that it was truly His body, complete with the wounds from His crucifixion.

Consider: In the garden outside the tomb, Mary Magdalene sees Jesus and mistakes Him for the gardener: "She turned around and saw Jesus standing there, but she did not recognize that it was Jesus." (John 20:14). On the road to Emmaus, Cleopas and another disciple walk and talk with Jesus, yet do not recognize Him until He breaks bread with them: "But their eyes were kept from recognizing him." (Luke 24:16). By the Sea of Tiberias, Jesus appears to seven disciples, speaks with them, and even gives fishing instructions, but again, they do not know it is Him until a moment of recognition dawns: "Then the disciple whom Jesus loved said to Peter, 'It is the Lord!'" (John 21:7).

This difficulty in recognizing Jesus suggests that His body, though still bearing the marks of the crucifixion, had undergone a mysterious transformation—something glorified, spiritualized, and elevated beyond the ordinary mortal state. His resurrection was not a mere resuscitation, but a passage into glorified existence.

The idea of a transformed, radiant body is consistent with Jewish eschatological hope, as prophesied by Daniel:

> Many of those who sleep in the dust of the earth shall awake, some to everlasting life, others to reproach and everlasting shame. The wise shall shine brightly like the splendor of the firmament, and those who lead the many to justice shall be like the stars forever. (Daniel 12:2–3)

Jesus Himself alluded to this prophecy when explaining the parable of the weeds to His disciples: "Then the righteous will shine like the sun in the kingdom of their Father." (Matthew 13:43)

The disciples had previously seen Jesus raise the dead—such as Lazarus or Jairus's daughter—where the person returned to life exactly as before. But Jesus' own resurrection was different: His body was transformed, sometimes unrecognizable, able to appear suddenly, yet still bore the wounds of the crucifixion. Though the disciples did not fully understand this change[143], they honestly recorded it, showing they were reporting a real, unexpected event, not inventing a story that fit their prior expectations. This unique transformation points to a new kind of glorified life, not just a return from death.

Another element that strongly reinforces the credibility of the Gospel accounts of the resurrection is the central role of women as the first witnesses to this monumental event. To modern readers, this might seem

---

[143] In 1 Corinthians 15, Paul explains that in the resurrection, our bodies will be imperishable (v. 42), glorious (v. 43), strong and powerful (v. 43), and spiritual or perfected (v. 44).

natural or unremarkable—but in the social and religious context of first-century Judaism, it was anything but.

In Jewish society at the time, women held a subordinate legal and cultural status. A woman was not permitted to speak with men in public and had to veil her face whenever she left her home. If a woman appeared unveiled in public, it was grounds for divorce. Women typically remained in the background—caring for the home, raising children, and serving under the authority of their husbands. When guests arrived, women ate separately, often in another room.

Marriages were generally arranged by parents, and a woman's greatest hope was not personal fulfillment, but that her husband might treat her better than her father had. In the synagogue, women were restricted to separate areas, and they were forbidden to read aloud from the Scriptures. A well-known rabbi of the time, Rabbi Eliezer, went as far as to say: "It would be better for the words of the Torah to be burned than entrusted to a woman."

Women were not permitted to recite prayers like the Shema[144], nor lead blessings at meals. Most striking of all, a woman's testimony was not valid in court[145]—she could not be a legal witness under Jewish law.

Given this context, it would be an understatement to say that first-century Jewish society was patriarchal. It was a culture in which women were often dismissed, restricted, and marginalized.

This is why the resurrection accounts stand out so powerfully. If the evangelists had fabricated the story of the resurrection—to fulfill prophecy or bolster belief—they would not have chosen a woman, much less Mary Magdalene, as the first witness. Her testimony would have

---

[144] For the Jews of that time, the Shema held a place of central importance, much like the Our Father does for Christians today. The Shema is found in the biblical texts of Deuteronomy 6:4–9 and 11:13–21, as well as Numbers 15:37–41. It is a declaration of faith in the one God and a call to love and obey Him with all one's heart, soul, and strength.

[145] According to traditional interpretation, Sarah—Abraham's wife—lost credibility after she denied laughing when God foretold that she would bear a child in old age. The belief arose that if she could lie to God, she could lie to anyone. This moment, described in Genesis 18:1–15, contributed to a lasting perception that cast doubt on the trustworthiness of women.

been considered untrustworthy by society, and she herself was a controversial figure, described in the Gospels as the one: "from whom seven demons had gone out." (Luke 8:2).

Luke clearly distinguishes between women healed of illness and those delivered from demonic possession—and Mary is identified in the latter group. This suggests that her past was widely known and scandalous, even if she had since become a faithful follower of Christ.

So why would the evangelists record that Mary Magdalene was the first to encounter the risen Lord?

Because that is exactly what happened.

Had they invented the story, they would have selected more credible male witnesses—perhaps Joseph of Arimathea or Nicodemus, both respected members of the Sanhedrin, men of wealth and influence. But they did not. Instead, they faithfully recorded what would have been, in the eyes of their culture, a "problematic" and unconvincing account—unless, of course, it was true.

As I explained in the fourth thesis of the second chapter, when the evangelists wrote the Gospels, they frequently noted when an event occurred in fulfillment of Scripture. This pattern appears throughout their writings in phrases such as: "This happened so that the Scripture might be fulfilled...", "But this took place to fulfill what is written in the Law...", "Then what had been spoken by the prophet was fulfilled...", "All this happened so that the writings of the prophets might be fulfilled...", or simply, "As it is written..."

This literary and theological pattern shows that the evangelists were deeply aware of the messianic prophecies and eager to demonstrate how Jesus' life fulfilled them. Yet curiously, when it comes to the resurrection of the Lord, the Gospels do not consistently include these formulaic statements of prophetic fulfillment—even though Scripture had foretold it (e.g., Psalm 16:10; Isaiah 53:10–11; Hosea 6:2).

If the resurrection narratives had been fabricated or exaggerated in an effort to persuade skeptics, one would expect the writers to have explicitly connected the event to such prophecies. Yet they did not. Why?

We can only speculate. The magnitude of the event spoke for itself. The early witnesses were still processing its mystery, too overwhelmed to tie it immediately to prophecy in the same structured way. Or the evangelists wanted their readers to see the fulfillment for themselves, rather than be told. Whatever the reason, the absence of these formulaic prophetic references in the resurrection narratives underscores a key point: the evangelists were not crafting a cleverly engineered story. Rather, they were reporting what they saw with raw honesty and without embellishment.

Indeed, we have four distinct perspectives on the same event, written without a coordinated agenda, without literary polishing, and without trying to protect reputations. The Gospels do not present themselves as hero stories. There are no brave men, no wise disciples, no attempts to portray the apostles in a favorable light. On the contrary, the authors confess their own cowardice, abandonment, and failure to understand Jesus' mission—even after He had foretold it repeatedly.

They openly record that the women remained faithful, while the men fled or doubted. These are not the kinds of details one would expect in a fabricated religious account, especially in a culture where male honor and social standing were paramount.

These narratives bear the marks of eyewitness testimony—unvarnished, consistent in substance yet diverse in detail, and radically honest, even when the truth was unflattering. The evangelists wrote not to promote themselves, but to bear witness to what had truly happened, even at the cost of their own credibility and their lives.

THIRD THESIS: JESUS: THE SON OF GOD, EVIL OR MAD?

Throughout history, we find numerous figures who gave their lives for a cause they passionately believed in. Mahatma Gandhi dedicated his life—and lost it—in the struggle to liberate India from British rule through non-violent resistance. Gaius Julius Caesar, in seeking to dismantle the corrupt Roman Republic that enriched a privileged few at the expense of the provinces, strove to establish a new political order, one he believed would serve the people more justly. Martin Luther King Jr., deeply inspired by Gandhi, led the American civil rights movement with a message of peaceful resistance and justice for African Americans. And in the early Church, Stephen, one of the first deacons, was stoned to death for preaching the Gospel of Christ.

All these individuals—and countless others—were driven by a profound conviction that their lives, even their deaths, could advance a cause that would outlast them. They were leaders, reformers, visionaries, and martyrs. They were remembered not because they achieved all their goals in life, but because their ideals eventually triumphed, often after their deaths.

Jesus of Nazareth is frequently placed among such historical figures—one more noble man who died for what He believed. But there is a critical difference that sets Him apart entirely.

Jesus did not die simply for a moral cause, or for justice, or peace, or reform. He died because He claimed to be someone altogether unique—the Son of God. No other religious leader in history made such a claim.

Take Muhammad, for example. He taught that the Archangel Gabriel had visited him to reveal the Qur'an, and he proclaimed himself to be a prophet, a chosen messenger, but not divine.

The Buddha also denied divinity. In one of the earliest recorded dialogues after his enlightenment, he was questioned by a group of seekers who were struck by his presence. The exchange is revealing:

"Are you a god?"
"No," he replied.
"Are you the reincarnation of a god?"
"No."
"Are you a sorcerer?"
"No."
"Are you a wise man?"
"No."
"Then what are you?" they asked, bewildered.
"I am the one who is awake."

This profound answer helped define Buddhism's spiritual path, but it made no claim to divinity.

Confucius never claimed to be more than a teacher and moral philosopher. Moses, revered in Judaism and Christianity alike, was God's chosen servant, but never divine. Even Saint Paul, one of Christianity's greatest apostles and theologians, never claimed to be God—only a servant of Jesus Christ.

But Jesus Christ is different. He did not merely speak for God—He spoke as God. He declared: "Before Abraham came to be, I AM." (John 8:58), "I and the Father are one." (John 10:30)

These declarations were so radical that the religious leaders of His time accused Him of blasphemy and demanded His death. As Thomas Schultz rightly observed:

> None of the recognized religious leaders —not Confucius, not Moses, not Muhammad, not Buddha, not Paul—none of them have claimed to be God; the exception is Jesus Christ. Christ is the only religious leader who has ever claimed to be deity and the only person who has convinced much of the world that he is.

The Jewish people were raised with a singular and sacred conviction: that obedience to the Law was the only path to salvation. From childhood, a devout Jew was taught that strict adherence to the Torah—God's Law given through Moses—was the one and only way to please God and gain entrance into eternal life. There was no alternative. The Law was everything.

And then, Jesus of Nazareth delivered what was, spiritually speaking, an atomic shock to that deeply religious society. He said:

> Do not let your hearts be troubled. You have faith in God; have faith also in me. [...] I am the way, and the truth, and the life. No one comes to the Father except through me. (John 14:1,6)

Would it be possible? Belief in Jesus, rather than in the Law, as the path to heaven?

Jesus did not say He was a way among many. He claimed to be the only way—a bold, categorical statement that directly challenged centuries of sacred teaching. The Law, revered and upheld by prophets and martyrs, was not the way to the Father, He said. He was.

For a devout Jew, this was almost unthinkable. The Books of the Maccabees, for example, recount the bravery of countless Jewish martyrs who endured torture and death rather than violate the Law. Their unwavering faith was in obedience—in keeping God's commandments at all costs. And now Jesus says that He, not the Law, is the way to salvation.

All the prophets and holy men of the Old Testament had urged the people to return to God with sincerity, to obey His commandments, and to listen to His voice. But neither of them ever claimed divinity or offered themselves as the path to salvation. Their mission was to point away from themselves and toward the Father.

Consider John the Baptist, the final prophet before Christ. When the people questioned whether he might be the Messiah, he refused the title clearly and humbly: "He admitted the truth and did not deny it. He declared, 'I am not the Messiah.'" (John 1:20)

He called the people to repentance, to conversion of heart, and to renewed fidelity to God. But he never claimed to save anyone. He knew full well that such a claim would be blasphemy, punishable by death.

And yet Jesus did make that claim—not subtly, but openly and repeatedly. On several occasions, He equated Himself with God the Father, with Yahweh, the Creator of all. "The Father and I are one." (John 10:30)

The Greek word used here for "one" is *hen*, in the neuter form, not the masculine (*heis*). This grammatical choice is essential: it indicates that Jesus and the Father are not the same person, but that they share the same divine nature or essence. Jesus was not merely aligning Himself with God's will—He was declaring ontological unity with God.

The religious leaders understood this perfectly. They were not confused or uncertain about what He meant. His claim to divinity enraged them, particularly because He did so publicly and boldly, even while breaking their legal traditions. For example, when Jesus healed a paralytic on the Sabbath, John recounts:

This was why the Jews were all the more determined to kill him: not only was He breaking the Sabbath, but He was also calling

God his own Father, thereby making himself equal to God. (John 5:18)

There is no doubt that both Jesus and His Jewish audience clearly understood the meaning and implications of His words. This was not a parable, nor a veiled metaphor. Jesus was making an explicit and undeniable claim to divinity.

One of the most direct examples of this occurs in John 8:58, where Jesus declares: "Amen, amen, I say to you: before Abraham came to be, I AM."

This declaration is remarkable for two reasons.

First, Jesus introduces it with the solemn formula "Amen, amen" (often translated "Truly, truly" or "Most certainly"), a strong and authoritative affirmation that signaled to His listeners the absolute truth of what followed. It was a rabbinic expression used to emphasize that the statement was not just opinion—it was a solemn, undeniable truth.

Second—and most dramatically—He refers to Himself as "I AM" (*Ego eimi* in Greek), a direct reference to the sacred name of God revealed to Moses in the burning bush: "God said to Moses: I AM WHO I AM. [...] This is what you are to say to the Israelites: I AM has sent me to you." (Exodus 3:14)

By taking this divine name upon Himself, Jesus was appropriating the incommunicable and unutterable name of the Creator (see Appendix A). And if any people on earth understood the gravity of invoking that name, it was the Jews—and Jesus Himself was one of them. This was not accidental or poetic—it was intentional and theological.

As if claiming to be the Messiah were not already controversial, Jesus went even further. He demanded the same honor that was due to God the Father:

The Father judges no one but has entrusted all judgment to the Son, so that all may honor the Son just as they honor the Father. Whoever refuses to honor the Son refuses to honor the Father who sent him. (John 5:22–23)

Before Jesus, no one—neither in the Old Testament nor in any known historical record—had dared to call God *Abba*. Jewish prayer traditions typically opened with the word Abhinu, meaning "Our Father", a reverent address that expressed a plea for mercy and forgiveness. It was respectful and formal, reflecting the awesome transcendence of the Creator.

But *Abba* was something entirely different.

Abba was the term used within the intimacy of family life, akin to "Papa", "Daddy", or "Papi". It was the most affectionate and personal form of address a child could use for his father. Not even King David, known for his deep relationship with God, dared use such familiarity. In Psalm 103, he writes: "As a father has compassion on his children, so the Lord has compassion on those who fear him." (Psalm 103:13)

Yet Jesus broke this precedent. In the Garden of Gethsemane, amid anguish and anticipation of His Passion, He prayed with unprecedented intimacy: "Abba, Father, all things are possible for you. Take this cup away from me. Yet not what I will but what you will." (Mark 14:36)

By addressing God as *Abba*, Jesus revealed the depth of His relationship with the Father, a relationship rooted not in fear, but in unity, love, and divine sonship.

Later, when Jesus stood before the Sanhedrin, He initially remained silent under questioning. But when the high priest pressed Him directly: "Are you the Messiah, the Son of the Blessed One?" (Mark 14:61). Jesus broke His silence and replied with divine clarity: "I am. And you will see the Son of Man seated at the right hand of the Power and coming with the clouds of heaven." (Mark 14:62),

Here, Jesus affirmed both His messianic identity and His divine authority using titles already charged with Old Testament weight: "Messiah" and "Son of Man"—titles used by prophets to describe the One sent from God, even God Himself in human form.

This is why Caiaphas tore his garments in horror and cried: "You have heard the blasphemy. What is your verdict?" (Matthew 26:65). To the high priest and the council, Jesus' words were not misunderstood—they were blasphemy, unless they were true.

And this is the heart of the dilemma. If Jesus was not who He claimed to be, then He was not merely a misguided teacher—He was either delusional, or deceitful on a grand scale. He told people that faith in Him was necessary for salvation, that He alone could forgive sins, and that He and the Father were one.

When Jesus forgave sins, He did not act as a priest, interceding between sinner and God. He spoke as the One offended, with complete authority: "But that you may know that the Son of Man has authority on earth to forgive sins"—he said to the paralyzed man—"I say to you, get up, pick up your mat, and go home." (Luke 5:24). On another occasion, He went even further—pronouncing both forgiveness and salvation over a woman of ill repute: "Your sins are forgiven." (Luke 7:48). The onlookers were scandalized: "Who is this who even forgives sins?" (Luke 7:49). But Jesus reaffirmed His verdict: "Your faith has saved you; go in peace." (Luke 7:50).

This is not the behavior of a mere teacher. It is the behavior of someone who consciously claimed to be God.

The great Christian thinker C.S. Lewis, in his classic *Mere Christianity*, addresses this exact point:

> I am trying here to prevent anyone from saying the really foolish thing that people often say about Him: 'I'm ready to accept Jesus as a great moral teacher, but I don't accept His claim to be God.' That is the one thing we must not say. A man who was merely a man and said the sort of things Jesus said would not be a great moral teacher.
> He would either be a lunatic—on the level with the man who says he is a poached egg—or else he would be the Devil of Hell. You must make your choice. Either this man was, and is, the Son of God—or else a madman or something worse."
> "You can shut Him up for a fool, you can spit at Him and kill Him as a demon—or you can fall at His feet and call Him Lord and God. But let us not come with any patronizing nonsense about His being a great human teacher. He has not left that open to us. He did not intend to.

In the field of information systems, we frequently use tools known as decision trees—logical diagrams that map out actions and outcomes based on a sequence of questions and answers. They help us visualize the process of rational decision-making, especially when multiple possibilities are at play.

This same model can be applied to Jesus' claim to be God. When He made this extraordinary assertion, only two fundamental options exist:

- His claim is true.
- His claim is false.

If the claim is false, there are only two further possibilities:

- Jesus knew His claim was false.
- He did not know it was false.

If He did not know, then He was deluded—a lunatic—mentally unstable, yet paradoxically preaching the most morally profound teachings in human history. If, on the other hand, He knew His claim was false, then He was deliberately deceiving others—a liar, a hypocrite, and a manipulator who encouraged people to stake their eternal souls on a lie. Worse still, He would be a narcissist so committed to His delusion that He willingly accepted torture and death for it. Such a person would be not only wicked but also utterly irrational.

But if His claim is true, then Jesus is exactly who He said He is: the Lord, the Messiah, the Son of God, God made flesh.

The evidence—historical, textual, logical, and spiritual—points consistently to the truth of Jesus' claim. But many still reject it—not because of flaws in the evidence, but because of the moral implications that acceptance would entail. To recognize Jesus as Lord means acknowledging His authority over one's life, submitting to His teachings, and accepting His call to repentance and discipleship.

This is why the question of Jesus' identity remains the most important decision any person can make. The title of this argument presents three alternatives:

Jesus was either a liar, a lunatic, or the Son of God.

Review the evidence. Consider the logic. Reflect with intellectual integrity and moral honesty.

Now make your choice.

FOURTH THESIS: THE SCENE OF THE EVENTS

The first element of this sorrowful scene is a corpse: Jesus died on the cross. Some opponents of the resurrection argued that the Master did not actually die but merely survived the crucifixion and was taken down from the cross while still alive.

However, a significant medical study challenges this notion. Dr. William Edwards, Dr. Wesley Gabel, and Dr. Floyd Hosner, pathologists from the Mayo Clinic in Rochester, Minnesota, published a detailed report on the physical death of Jesus. Their findings appeared in the *Journal of the American Medical Association*, in the issue dated March 21, 1986 (recap):

> Let us first consider the physical condition of Jesus. The demands of His ministry, including extensive travel on foot across the land of Israel, would have been impossible without robust health. We can reasonably assume that Jesus was in excellent physical condition prior to His arrest in the Garden of Gethsemane.
> Following His arrest, however, a cascade of physical and emotional trauma began: emotional stress, sleep deprivation, lack of food and water, severe beatings, and the long walk to Golgotha all made Jesus more vulnerable to the devastating physiological effects of Roman scourging.
> The Gospels report that in Gethsemane, Jesus experienced such agony that He sweated blood—a phenomenon known today as hematohidrosis (bloody sweat) (cf. Matthew 26:36–38; Luke 22:44). Science identifies this condition as a rare, stress-induced hemorrhaging of the sweat glands, which leaves the skin extremely fragile.
> According to the medical report, during the scourging, Jesus suffered deep lacerations inflicted by a flagrum— a whip consisting of multiple leather thongs with metal balls and bone fragments at their ends (cf. Matthew 27:24–26). These whips wrapped around the victim's torso, tearing into the subcutaneous tissue and even skeletal muscle, inflicting wounds so severe that the body was often left on the verge of circulatory collapse or death.
> The volume of blood loss during scourging was a determining factor in how long a victim might survive on the cross. In Jesus'

case, the extent of the blood loss likely brought Him into a state of hypovolemic shock—a condition where blood volume is critically low, reducing the heart's ability to pump effectively.

Adding to the trauma, the Gospel of Matthew recounts that Jesus' wounded back was covered with a cloak by the soldiers, only to be ripped off later, reopening and aggravating His injuries (cf. Matthew 27:27–31).

At the crucifixion site, Jesus' arms and legs were fully stretched. Nails were driven between the radius and carpal bones of the wrists. Though no bones were broken, the periosteum—a highly sensitive membrane covering the bones—was likely pierced, causing intense pain. The nails probably severed the median nerve, producing fiery nerve pain in both arms and rendering part of His hands paralyzed, causing a "claw-like" hand deformity.

The nails in His feet likely pierced His tarsal bones, also injuring major nerves and contributing further to His pain. But the most critical effect of crucifixion was on breathing. The body was fixed in a position that made exhalation extremely difficult, resulting in shallow breathing, muscle cramps, and progressive asphyxiation.

According to the Gospel of John, when a Roman soldier pierced Jesus' side with a spear, a sudden flow of blood and water was observed (cf. John 19:34). From a modern medical perspective, this indicates that the spear likely penetrated the right lung, the pericardium, and the heart, ensuring death.

Considering these findings, the suggestion that Jesus merely survived the crucifixion is incompatible with modern medical knowledge. The physiological evidence described by Drs. Edwards, Gabel, and Hosmer, and affirmed by the Gospel witnesses, confirms that Jesus was indeed dead when taken down from the cross

Roman soldiers were so accustomed to death that they could easily recognize it. They knew, beyond doubt, when someone had died. This explains the reaction of the Roman centurion standing before Jesus at the moment of His death: "Truly, this man was the Son of God." (Mark 15:39)

It was likely this same soldier who later confirmed Jesus' death to Pontius Pilate, allowing him to release the body to Joseph of Arimathea when he came to request it:

Pilate was surprised to hear that He was already dead, and he summoned the centurion and asked whether Jesus had already died. When the centurion confirmed this, Pilate granted the body to Joseph. (Mark 15:44–45)

The second element in this scene is the tomb where the body of Jesus was placed late on that Friday afternoon. The word "tomb" appears thirty-two times in the biblical accounts of the resurrection—underscoring the central importance the apostles placed on this location.

The Church historian Eusebius of Caesarea, known as the "Father of Church History," recorded in his work *Theophany* a description of the tomb as relayed to him by Empress Helena, the first imperial patron of the Holy Sepulchre:

> The tomb itself was a cave that had been carved out; a cave that had been cut into the rock and had not been used by anyone else. It was necessary that the tomb, which in itself was a marvel, cared only for a corpse.

In March 2016, the six Christian orders that share custodianship of the Holy Sepulchre—the Greek Orthodox, Roman Catholic, Armenian Apostolic, Syrian Orthodox of Antioch, Coptic, and Ethiopian Churches—granted permission to a team from the National Technical University of Athens to inspect and restore the Edicule, the structure that covers the tomb.

The restoration project, which cost over $4 million, was financed in large part by King Abdullah II of Jordan and a $1.3 million donation from Mica Ertegun[146] to the World Monuments Fund. While the archaeologists concluded that it is not possible to verify with absolute certainty that the current site is the exact location of Christ's burial, they affirmed that the Church of the Holy Sepulchre and the tomb it encloses occupy the same area identified in the fourth century by Saint Helena and her son, Emperor Constantine.

---

[146]She is the widow of Ahmet Ertegun, a Turkish-American businessman and music producer. Ertegun was best known as the co-founder and president of Atlantic Records, the influential record label that helped launch the careers of iconic artists such as Ray Charles, Led Zeppelin, Phil Collins, and Crosby, Stills, Nash & Young.

When Helena and her entourage arrived in Jerusalem around AD 325, their search led them to a Roman pagan temple, built approximately two centuries earlier. The structure was dismantled, and beneath it they discovered a rock-cut tomb within a limestone cave. To reveal the burial chamber, where the body of Jesus had been laid on a stone bench, the upper portion of the cave was removed. To preserve this sacred place, the Edicule—a small shrine-like structure—was built over the tomb. That structure, with many restorations, remains standing to this day.

During the 2016 restoration, samples of mortar from the Edicule were extracted and dated in two independent laboratories. The analysis confirmed that the construction materials dated back to the fourth century. This finding supports the continuity of the site, indicating that the location venerated today as the tomb of Jesus has remained the same for over 1,700 years, despite enduring numerous attacks, collapses, and reconstructions throughout its long history.

The third element of the scene is the grave. Remarkably, we know more about the burial of Jesus than we do about the burials of any other prominent figures of antiquity—including pharaohs, kings, emperors, and philosophers.

We know who took possession of Jesus' body after His death was officially confirmed. We know the name of the man who donated the embalming spices, as well as the amount he donated. The individuals involved in the burial preparations, following the customary rites of the time, are also recorded by name.

We know who owned the tomb, including his place of origin, religious affiliation, economic status, and profession. We know the precise location of the tomb, and we are told how many times it has been used previously. We even know the material from which it was made.

The approximate day and time that the body was laid in the tomb are preserved in the Gospel record. We know how the tomb was sealed, and who was assigned to guard it for three days.

Such a level of historical detail regarding a burial is unmatched in the ancient world. No other figure from antiquity, regardless of status or legacy, has had the circumstances of their burial preserved with such specificity.

The fourth element of the scene is the stone. It is known that the stone was round, large, and extremely heavy. This explains the concern of the women who, on their way to the tomb, wondered who would roll it away.

The tomb could be entered upright, without the need to bend down, suggesting that the stone had a diameter of approximately five feet, or more. Based on this size, its thickness would have been at least twelve inches. With these dimensions, the stone would have weighed more than two tons—an object far too heavy to move without significant effort.

This assessment aligns with the Gospel descriptions: "And he rolled a large stone across the entrance to the tomb." (Matthew 27:60) and "They looked up and saw that the stone, although it was extremely large, had already been rolled back." (Mark 16:4)

The fifth element in the scene is the seal. I will dedicate the following thesis exclusively to this subject, as it holds great importance in understanding the full context of the burial and resurrection narrative.

The sixth element of the scene is the guard. Because Jesus had repeatedly announced that He would rise from the dead on the third day, the Sanhedrin feared that His disciples might attempt to steal the body. If the corpse were to disappear, they reasoned, the disciples could then proclaim the much-anticipated resurrection of the one who had claimed to be the Son of God.

For this reason, the Sanhedrin succeeded in convincing Pontius Pilate to assign a "guard troop" to monitor the tomb—meaning Roman soldiers. Given the concern that all twelve apostles, or at least the remaining eleven, might attempt such a theft, the number of soldiers had to be proportional to the perceived threat.

A precedent for such proportional guarding is found in the Acts of the Apostles, when King Herod placed Peter under arrest and had him guarded by sixteen soldiers:

About that time, King Herod began a persecution of certain members of the Church. He had James, the brother of John, put to death by the sword. Seeing that this pleased the Jews, he proceeded to arrest Peter also. [...] After arresting him, he put him in prison and assigned four squads of four soldiers each to guard him. (Acts 12:1–4)

It is reasonable, then, to assume that a similar number of soldiers— between four and sixteen—was assigned to guard the tomb of Jesus.

The *Strategikon*, a Roman military manual, describes the punishment for a soldier who fell asleep on watch: a brutal form of discipline called *animadversio fustium*, in which the offender was publicly beaten with rods until he lost consciousness. This harsh reality helps explain why the jailer of Paul and Silas, believing they had escaped after an earthquake, drew his sword to take his own life:

When the jailer was roused from his sleep and saw the doors of the prison wide open, he drew his sword and was about to kill himself, thinking that the prisoners had escaped. (Acts 16:27)

The historian Polybius[147] records that a guard troop typically consisted of four to sixteen men, who were relieved every eight hours. These Roman soldiers assigned to Jesus' tomb would have been fully aware of the consequences of failing in their duty.

Is it plausible to believe, then, that all of them fell asleep simultaneously, and that no one awoke while the disciples rolled away a massive stone and removed the body?

The whole scene of the burial place of Jesus has enormous historical support. Never has a criminal produced so much concern after his execution. Above all, someone condemned to die on the cross had never had the honor of being guarded by a squad of soldiers. All judicial and police measures at the time, in addition to those dictated by prudence,

---

[147]Polybius (Megalopolis, Greece, BC 200 – 118) was a Greek historian considered one of the most important figures in the field of historiography. He is credited with writing the first true universal history, aiming to explain how Roman hegemony came to dominate the Mediterranean world. To achieve this, Polybius demonstrated how political and military events across various regions were interconnected, presenting a cohesive and comprehensive account of the rise of Rome.

were taken to prevent the corpse of Jesus from moving even one inch from the place where it had been deposited that Friday. Even so, three days later, the body was gone.

Today we can feel with our own hands the rock of the place where Jesus was enshrouded and touch the stone on which his body rested in that tomb, which is still empty.

FIFTH THESIS: THE SEAL

The Gospel of Matthew records a key detail regarding the burial of Jesus: "So they went and secured the tomb by placing a seal on the stone and setting a guard." (Matthew 27:66).

This detail is often overlooked, yet it is extremely significant. A casual reading may suggest that the word "seal" simply refers to the stone placed over the entrance to the tomb—which, of course, is true in a physical sense. However, that is not what the evangelist is emphasizing.

To understand this better, we can turn to a passage in the Book of the Prophet Daniel, which offers a historical precedent:

> A stone was brought and placed over the mouth of the den. The king sealed it with his own signet ring and with those of his nobles, so that nothing could be altered with regard to Daniel. (Daniel 6:17)

In this context, the seal was a legal and official measure. It involved a rope or cord stretched across the stone that blocked the entrance. The ends of this cord were fastened to the rock wall using soft clay, and a seal was impressed into the clay using the signet ring of a Roman authority—in this case, Pontius Pilate, or someone acting under his authority.

The impression of the ring served as formal Roman authentication, making it clear that the tomb was under official watch. To move the stone, the seal would first have to be broken, which would constitute a direct violation of Roman authority.

Breaking such a seal was not a matter to be settled with the Sanhedrin or any local Jewish court. It was a serious offense against Rome itself—an act of defiance punishable by severe consequences.

This physical sealing technique remained in use until the late 17th century, when materials like sealing wax began to be employed to secure royal correspondence and other official documents. Typically red in color, sealing wax was made from a mixture of rosin, shellac, turpentine, and vermilion. Once a document was prepared, a small amount of this mixture was melted at the closure, and a government seal or the king's signet ring was pressed into the wax and left to harden.

Once dry, the document could only be opened by breaking the seal—a deliberate act that would signal tampering. This method guaranteed privacy and protected the integrity of the message inside.

Why, then, did Governor Pilate go to such lengths to protect Jesus' tomb with such official care?

To answer that, we must rewind the timeline by a few hours to the interrogation of Jesus before Pilate. According to the Gospel of John, during the trial, the crowd insisted that Jesus should be crucified because He had "claimed to be the Son of God": "We have a law, and according to that law He ought to die, because He claimed to be the Son of God." (John 19:7). Upon hearing this, the Roman governor became more afraid: "When Pilate heard what they said, he was even more afraid." (John 19:8).

Like many Romans, Pilate was deeply superstitious. The possibility that Jesus possessed divine powers, or that He was a god or a descendant of the gods who had taken human form (cf. Acts 14:11), filled him with dread. If such were the case, then Pilate had just permitted the scourging and humiliation of a supernatural being—who might well seek revenge.

His fear was intensified by his wife's dream, a troubling vision that prompted her to warn him: "Have nothing to do with that righteous man, for today I have suffered much in a dream because of him." (Matthew 27:19). Driven by anxiety, Pilate questioned Jesus privately, asking: "Where do you come from?" (John 19:9).

Pilate was not inquiring about Jesus' geographical origin, since he already knew He was a Galilean (cf. Luke 23:5–7). What Pilate sought to understand was Jesus' nature. The question was prompted by what Jesus had previously told him: "My kingdom does not belong to this world." (John 18:36).

Pilate wanted to know: Did this man belong to the realm of mortals, or was He from the realm of the gods?

But the few words Jesus offered in response did nothing to calm the governor's fear. In the end, Pilate chose to appease the crowd, ordering the death of a man he still suspected might possess divine power. Better, he reasoned, to please the people than to risk a public uprising—especially during Passover, when Jerusalem was filled with pilgrims and religious fervor.

If Pilate had been deeply unsettled by Jesus' words during the trial, that anxiety was nothing compared to what followed. The Synoptic Gospels recount an extraordinary event: "From noon until three o'clock in the afternoon, darkness came over the entire land." (Matthew 27:45; Mark 15:33; Luke 23:44).

Was this darkness the result of a solar eclipse, as some have speculated?

The answer is no. A solar eclipse cannot account for darkness lasting more than three hours. In fact, according to astronomical calculations, the maximum duration of a total solar eclipse is just seven minutes and thirty-one seconds. Nevertheless, some Bible translations—such as *The Book of God's People*[148]—refer to it as an eclipse.

Ancient historians, including Sextus Julius Africanus[149] and Tertullian[150], offered natural explanations such as a *chamsin* (a fierce sandstorm) or dense black clouds forewarning a severe storm. Regardless of the precise cause, the event only deepened Pilate's

---

[148]"It was about noon, and darkness came over the whole land until three in the afternoon, because the sun was eclipsed. The veil of the temple was torn in two." (Luke 23:44–45)

[149]Sextus Julius Africanus (c. AD 160 – c. 240) was a Christian historian and apologist of North African origin, influenced by Hellenistic culture. He is regarded as the father of Christian chronology for his efforts to systematically date biblical and historical events.

[150]Quintus Septimius Florens Tertullian (c. AD 160 – c. 220) was a Church Father and a prolific writer active during the latter half of the 2nd century and the early part of the 3rd century.

superstitious fear. He must have longed for the day to end and to leave behind the disturbing chain of events unfolding before him.

Then, around three in the afternoon—the moment Jesus breathed His last—Pilate witnessed something even more terrifying: a powerful earthquake unlike anything he had ever experienced. "At that moment, the veil of the temple was torn in two from top to bottom, the earth shook, and rocks were split." (Matthew 27:51; cf. Mark 15:38; Luke 23:45)

In that moment, the Roman governor had no doubts left. He realized he had not only condemned an innocent man, but someone extraordinary—someone backed by a supernatural force.

Sextus Julius Africanus also recorded the unusual natural phenomena surrounding the death of Jesus. In the third book of his *Chronicle*, a work composed of five volumes, he wrote:

> A terrifying darkness fell over the whole universe; an earthquake broke the rocks; most [of the houses] of Judea and the rest of the land were razed to the ground. Thallus, in the third book of his *Histories*, regards this darkness as an eclipse of the sun, but without reason, it seems to me. [151]

In more recent times, geologists Jefferson B. Williams, Markus J. Schwab, and A. Brauer studied sediment disturbances in the Dead Sea region, near the shores of Galilee. Their research identified evidence of two major earthquakes: one significant quake around BC 31, and another, less intense but still notable, sometime between AD 26 and 36.

Their full study was published in *Geology Review*, Volume 54 (2012). While the authors were not entirely certain that the second quake could fully account for the tearing of the temple's sanctuary veil, they acknowledged that the margins of error regarding both the magnitude and exact date could allow for that possibility.

---

[151]Thallus was an early historian who wrote in Koine Greek a three-volume *History of the Mediterranean World*, covering events from before the Trojan War up to the 167th Olympiad.

Amid all these extraordinary events, Pilate's concern intensified—particularly over the possibility that Jesus' body might disappear, just as the Jewish leaders feared:

> Sir, we recall that while He was still alive, that deceiver said, 'After three days I will rise again.' Therefore, command that the tomb be kept secure until the third day, lest his disciples come and steal him away and tell the people, 'He has been raised from the dead.' This last deception would be worse than the first. (Matthew 27:63–64)

In response, all available legal and security measures were employed to guard the tomb, ensuring that no one would dare to disturb the body during those long three days.

SIXTH THESIS: THE EMPTY TOMB

Deciphering the mystery of the empty tomb and the post-resurrection appearances is key to demonstrating that the resurrection of Jesus was a historical event. As the Apostles' Creed affirms, "He was crucified, died, and was buried," and three days later, His body was no longer in the tomb.

What happened? Where is the body?

How can we explain that, despite all the measures taken by the authorities to protect the tomb, its contents disappeared on the third day?

The New Testament records multiple appearances of the risen Lord following His departure from the grave. Some of these were personal encounters—with Peter, Mary Magdalene, James, and very likely with His mother. Others were public, including one before a group of more than five hundred followers:

> Afterward, He appeared to more than five hundred brothers at one time, most of whom are still alive, although some have fallen asleep. (1 Corinthians 15:6)

The Bible is not the only source that mentions witnesses to the resurrected Christ. Several ancient historians also refer to these events, including Flavius Josephus in his *Antiquities of the Jews*, Cornelius Tacitus in *Annals*, and Gaius Plinius Caecilius Secundus (Pliny the Younger) in his letters to Emperor Trajan, among others[152].

The witnesses did not describe vague "sightings" or abstract visions. They spoke of encounters—of interactions with someone they recognized as real and alive, someone who spoke, walked, and ate with them.

- In the first appearance, to Mary Magdalene (John 20:11–18), Jesus spoke with her by name.
- In the second, to a group of women (Matthew 28:8–10), they spoke with Him and even clung to His feet.
- In the third, on the road to Emmaus (Luke 24:13–33), He walked with two disciples, explained the Scriptures, and shared a meal with them.
- In the fourth, with ten of the apostles gathered (John 20:19–22), Jesus showed them His wounds and ate with them.
- In the fifth, He appeared again to the apostles—this time with Thomas present (John 20:26–29). Jesus invited Thomas to touch His hands and place his hand into His side.
- The sixth appearance took place by the Sea of Galilee, where seven disciples were fishing. Jesus joined them and shared a meal.

Though these appearances were of a "transformed" body—a mystery that took the disciples time to fully grasp (cf. 1 Corinthians 15:38–57)—it was still clearly a real, physical body. He spoke, reasoned, walked, and ate. It was the body of a living person.

Throughout history, all kinds of theories have been proposed to explain the empty tomb—ranging from the fanciful to the more seemingly plausible. However, those who seek to dismiss the miracle of the resurrection are confronted with the challenge of fitting all the available evidence into their alternative explanations.

It is essential to emphasize the word "all", because any theory that accounts for only part of the evidence cannot be stronger than one that

---

[152]In the previous chapter, I included the corresponding bibliography for these historians and others.

explains the entire historical record. A credible hypothesis must encompass every relevant fact to be considered serious.

Let us examine some of the more popular non-resurrection theories that have attempted to explain the empty tomb:

- **The Catalepsy Theory**: Popularized by a heterodox Muslim group known as the Ahmadiyya Community, this theory claims that Jesus did not actually die on the cross but fell into a state of catalepsy, later waking up inside the tomb. According to this view, He exited the tomb by His own means and reunited with His disciples. However, this theory is contradicted by the position and condition of the burial cloths, as described by the beloved disciple—a detail I will explore in a later section. It is also logistically implausible: a severely wounded man could not have moved a two-ton stone, let alone from the inside. If this had occurred, the Roman guards would not have run to the chief priests to request help fabricating an explanation to avoid the consequences they faced for allowing the body to disappear. Furthermore, as discussed in the fourth thesis of this chapter, a detailed medical report confirms that Jesus died on the cross.

- **The Hallucination Theory**: Proposed by the French theologian and orientalist Joseph Ernest Renan in the late 19th century, this theory asserts that the post-resurrection appearances of Jesus were hallucinations, brought on by the disciples' overwhelming emotional trauma.
  However, as previously noted, the Gospel accounts describe not mere "sightings" but interactions—spoken conversations, physical contact, shared meals. The witnesses testified not to visions, but to encounters with someone alive, tangible, and reasoning.

- **The Wrong Tomb Theory**: This idea, advanced by Kirsopp Lake, a professor of New Testament at the University of Oxford in the mid-20th century, claims that the women mistakenly went to a different, empty tomb.
  This theory fails to account for the numerous post-resurrection appearances of Jesus. The Gospels clearly show that the disciples

themselves initially thought they were seeing a ghost[153], but Jesus dispelled that idea: "While they were still speaking about these things, Jesus himself stood in their midst and said to them, 'Peace be with you!' In their panic and fright, they thought they were seeing a ghost. But He said to them, 'Why are you disturbed? Why do such doubts arise in your hearts? Look at my hands and my feet, that it is I myself. Touch me and see; for a ghost does not have flesh and bones as you see I have.'" (Luke 24:36–39). Jesus invited them to touch Him—He wanted to make clear that He was not a spirit, but present in body and soul. He even asked them for something to eat, and they gave Him a piece of roasted fish, which He ate in their presence. The text leaves no doubt: what they saw was not a ghost, but Jesus in His resurrected, still-wounded body.

Another point that undermines the "wrong tomb" theory is that the women knew the tomb well. The Gospels affirm their presence at the burial: "The women who had come with Jesus from Galilee followed and saw the tomb and how his body was laid in it." (Luke 23:55) or "Mary Magdalene and the other Mary remained sitting there opposite the tomb." (Matthew 27:61)

- **The Qur'an's Version**: Islam offers its own version of events regarding Jesus' death and the empty tomb. In the Qur'an, Jesus is known as '*Isa ibn Maryam*', meaning "Jesus, son of Mary." The story begins with his grandmother Ana, who dedicated her daughter Mary to the service of the temple before her birth. Mary, shown to be deeply devoted to God, miraculously conceived Jesus while still a virgin, after being visited by an angel. According to the Qur'an, Jesus grew in wisdom, became a prophet, was called the Messiah, and performed many miracles. Eventually, He was flogged and sentenced to death by crucifixion—but He survived. His disciples secretly healed Him, and He escaped to continue His mission to the lost tribes of Israel.

---

[153]Some Bible translations use the words "ghost" or "spirit," though their meanings can differ significantly. The term "ghost" is often associated with demonic entities, whereas "spirit" can have a broader range of meanings. For example, expressions like "Spirit of the Lord" or "Holy Spirit" refer to God Himself. However, the term "spirit" can also appear in negative contexts, as in the phrase "unclean spirit," which refers to a demon.

Under the name *Yuz Asaf*, He traveled east, eventually reaching Kashmir, India, where He is said to have died at the age of 120. Today, a modest shrine in Srinagar, a city in northern India, marks what some believe to be His tomb. It receives few visitors.

In addition to the arguments above that challenge these theories, we must also consider that, after the resurrection, Jesus' body was no longer the same. It could pass through locked doors, something that deeply impressed the apostles. The Gospels tell us:

> In the evening on that same day, the first day of the week, *the doors were locked* in the room where the disciples were, for fear of the Jews. Jesus came and stood in their midst and said to them, 'Peace be with you!' (John 20:19, *emphasis* mine)

Lastly, we cannot forget that the disciples also witnessed His ascension:

> As He said this, He was lifted up while they looked on, and a cloud took him from their sight. While they were gazing up into the sky as He was going, suddenly two men dressed in white stood beside them and said, 'Men of Galilee, why are you standing here looking up into the sky? This Jesus who has been taken from you into heaven will return in the same way you saw him go to heaven.' (Acts 1:9–11)

It should be clear by now that these theories may account for some of the evidence—but not all. And as mentioned earlier, partial explanations cannot outweigh a theory that fits the full body of available facts.

To the list of alternate explanations for the resurrection, there is one more that deserves special attention—because it is found in the Gospels themselves:

> While the women were on their way, some members of the guard went into the city and reported to the chief priests all that had happened. After meeting with the elders and formulating a plan, they gave a large sum of money to the soldiers and instructed them, '*You are to say, "His disciples came during the night and stole him while we were asleep.*" If the governor hears of this, we will placate him and protect you from any trouble.' The

soldiers accepted the money and did as they were instructed. And this story is still told among the Jews to this very day. (Matthew 28:11–15, emphasis mine)

An empty tomb, in itself, does not conclusively prove that a resurrection occurred. But it does raise a profound question: Was the disappearance of Jesus' body the result of divine action or human interference?

The facts are simple. Jesus was buried, anointed, and wrapped in a linen shroud on Friday, before sundown, in a tomb donated by Joseph of Arimathea. When the women returned on Sunday morning to complete the burial rites—cut short due to the onset of the Sabbath—the body was no longer there.

This situation presents two explanations:

- Someone entered the tomb and removed the body — a human act, a theft.
- Jesus rose and left the tomb by His own power — a divine act, the resurrection.

If the first option is true—if it was a robbery—then the natural question follows: Who removed the body? Only two groups of people could be considered suspects: His friends or His enemies.

Throughout history, the desecration of graves has unfortunately been a common crime, and it continues in many parts of the world today. However, it is essential to recall that the tomb of Jesus was, in effect, Roman territory—under the authority and laws of the Empire. And Roman law severely punished those who tampered with graves. Offenders faced fines ranging from 100,000 to 200,000 sesterces, a staggering amount[154].

---

[154]*Actio de sepulchro violato* ("action for violation of a tomb") was a legal remedy in Roman law. The praetor granted this action against anyone who intentionally violated, inhabited, or built upon a tomb that did not belong to them. If the rightful holder brought the claim, the penalty was determined as *quanti ob eam rem aequum videbitur*— "as much as seems fair and equitable for the matter." However, if the rightful owner chose not to pursue the claim or if no owner could be identified, the praetor permitted a popular action as a subsidiary measure, imposing a fine of one hundred thousand sesterces for

To better understand the severity of this penalty, consider the following: According to Tacitus (*Annals,* Book I, chapters 17:4–5), Roman soldiers stationed along the Rhine were paid four sesterces per day—a wage that had to cover even their own uniforms. A British writing tablet dated AD 75 records the sale of a slave named Vegetus for 2,400 sesterces. In this context, a fine of 100,000 or 200,000 sesterces would have been economically devastating for any individual—making grave robbing a high-risk crime.

Moreover, this tomb was not just any tomb: it had been officially sealed by Roman authority. The seal bore the imperial insignia, and breaking it was equivalent to defying the emperor himself.

Who, then, would have dared to approach the stone—let alone move it?

SEVENTH THESIS: DID THE ENEMIES OF JESUS STEAL THE BODY?

The only explanation that Mary Magdalene could imagine upon finding the tomb empty was that Jesus' enemies had stolen the body:

> They asked her, 'Woman, why are you weeping?' She answered, 'They have taken my Lord, and I do not know where they have laid him.' (John 20:13)

However, logic tells us that the enemies of Jesus could not have been responsible for such a sacrilegious act. They were the ones most invested in preventing any rumor that the prophecy of the third-day resurrection might have come true. Let us recall the words of the chief priests to Pilate on Friday afternoon:

> Therefore, give orders that the tomb be made secure until the third day, lest his disciples come and steal him away and tell the people, 'He has been raised from the dead.' This last deception would be worse than the first. (Matthew 27:64)

---

violation and two hundred thousand sesterces for habitation or construction. (Source: Roman Law, Gumesindo Padilla Sahagún)

Thus, in addition to having the tomb sealed with the imperial seal, they convinced the governor to assign a squad of Roman soldiers to guard it.

Despite these precautions, when the body disappeared, the guards had no explanation. Fearing the consequences of failing in their duty, they went to the Sanhedrin for help in crafting a narrative that would protect them: "You are to say, 'His disciples came during the night and stole him while we were asleep.'" (Matthew 28:13)

This invented version presents a logical problem. It should not be assumed that all the soldiers fell asleep at the same time, or that guard duty was only carried out during the day. In concocting this false story, the soldiers were, in fact, incriminating themselves against a serious military offense—falling asleep on duty. As discussed in the fourth thesis of this chapter, the punishment for such a crime was to be flogged in public until unconsciousness.

Therefore, the bribe offered to them must have been substantial enough to justify the risk: "They gave a large sum of money to the soldiers." (Matthew 28:12)

Had the soldiers used the story for which they were paid, and if Pilate had found out, his wrath would have fallen upon them—especially because their account could not withstand even basic scrutiny.

- If all the guards had fallen asleep, how could they identify the culprits?
- If they knew it was the disciples, how could they also claim they had been asleep?
- And if they had not been asleep, how could they have allowed the theft to occur?

Whatever version they offered, they would end up in serious trouble. Their only hope was that their superiors would not investigate further, allowing them to preserve their lives—and enjoy the bribe they had accepted.

Several days after that resurrection Sunday, Peter gave his first public speech before a crowd of Jews and Gentiles. Standing in the public square, he interpreted the Scriptures (Acts of the Apostles 2:14–41), referenced the prophecies about the Messiah, and emphasized that it had been foretold that the Messiah would not remain in the grave, nor would His body undergo corruption. He then testified, along with the other apostles, that they were eyewitnesses to the resurrection of the Master.

The response to Peter's proclamation was overwhelming. The Scriptures record that on that day, "about three thousand persons" believed and were baptized.

Now, if the enemies of Jesus had possessed His corpse, would this not have been the perfect moment to discredit the disciples' claims of a resurrection? Would this not have been a golden opportunity to expose them as liars?

They could have simply produced the body, laid it in the middle of the square, and exposed the apostles as frauds, thereby extinguishing, at its very inception, the fledgling Church that was beginning to take form. After all, the resurrection of the Messiah was the foundation of the Church that, by the command of Jesus, had just begun to be built. If that claim were disproven, Christianity would have ended right then and there.

If Jesus' enemies had stolen His body from the tomb, then where did they put it? Could there have been a more secure location than the guarded tomb, sealed by the authority of Caesar and protected by soldiers of the most powerful army in the world?

If the enemies of Jesus had deliberately removed the body to fool the disciples, only to later reveal it in public, why did that moment never come? Why did they remain silent?

It becomes increasingly clear that the enemies of Jesus must be ruled out as suspects—because nothing harmed them more than the fact that the body had disappeared.

EIGHTH THESIS: DID JESUS' FRIENDS STEAL THE BODY?

The philosopher, politician, orator, and writer Lucius Annaeus Seneca, in his tragedy *Medea* (written in the year AD 56), penned the well-known phrase: *"Cui prodest scelus, is fecit"* — "He who benefits from the crime is the one who committed it."

Following this reasoning, the primary suspicion would naturally fall on the friends of Jesus, who would have stood to gain the most from the alleged theft of His body. But did they, in fact, commit such an act?

Given both the direct evidence and the circumstantial factors at our disposal, can we, without violating logic and reason, conclude that the disciples were responsible for removing the body of Jesus?

We are faced with two opposing camps:

- On one side, there were those who sentenced Jesus to death— namely, the Sanhedrin—a group with power, wealth, and an implicit alliance with the Roman government.
- On the other side were the disciples—a small group with no political influence, no legal recourse, and no protection from either civil or religious authorities.

With Jesus' execution, the Sanhedrin believed they had eliminated the source of the threat. Yet they knew there were still "seeds"—His followers—that might one day take root and spread. But could they eliminate the apostles? Did they have the means to prosecute them?

The answer is no.

The only charge that brought Jesus to death was blasphemy—His claim to be the Son of God. None of His disciples made such a claim. Therefore, there was no religious basis to accuse them of the same crime.

To hand them over to Roman justice, the Sanhedrin would have needed to charge them with a civil offense—something that violated Roman law. And they had the perfect candidate: if the disciples had broken the governor's seal and desecrated the tomb, this would have constituted a criminal offense, punishable under Roman authority.

As explained earlier, desecration of a tomb was considered a serious crime, and breaking the imperial seal without authorization was an offense that carried the maximum penalty. All the Sanhedrin had to do was provide evidence to Pilate that the disciples were guilty of this transgression, and the governor would have ensured their execution.

But that never happened.

Why?

Because no such evidence existed. And in the absence of proof, the Sanhedrin had no other choice but to bribe the guards, promising them protection in return for their silence and complicity:

> They gave a large sum of money to the soldiers and instructed them: 'You are to say, "His disciples came during the night and stole him while we were asleep." If the governor hears of this, we will placate him and protect you from any trouble.' (Matthew 28:12–14)

This fabricated story was their only option. It was a desperate solution born not of strength, but of lack of evidence.

NINTH THESIS: FROM COWARDS TO BRAVE?

As Zechariah prophesied ("Strike the shepherd, and the sheep will be scattered," cf. Zechariah 13:7), the apostles abandoned the Master at the time of His arrest and trial—an event carried out under questionable legality by temple officials and Roman soldiers, led to Him by Judas Iscariot.

Peter, more courageous than the others, followed at a distance, while most of the disciples went hiding. John eventually returned and remained with Jesus until His final moments at the cross. But even the one who had been called "The Rock"—on whom the nascent Church would be built—did not hold out long. After denying any association with Jesus, Peter withdrew, later weeping bitterly over his betrayal: "But Peter said, 'Woman, I do not know him.'" (Luke 22:57)

Eventually, Peter returned to the others. Yet fear gripped them all. They stayed behind locked doors until that first day of the week, when Jesus appeared to them in the upper room where they were hiding:

> On the evening of that same day, the first day of the week, the doors were locked in the place where the disciples were, for fear of the Jews. Jesus came and stood in their midst and said to them, 'Peace be with you.' (John 20:19)

Can we reasonably believe that these same men—who had shown themselves to be fearful and demoralized—suddenly found the strength to steal the body of their Teacher, their Friend, their Lord?

Would they have dared to face a Roman guard detachment, or had the audacity to break a seal that bore the signet of Caesar's representative?

On Friday, they had seen their beloved Teacher murdered in the most brutal way possible—the very man for whom they had left everything, with whom they had lived and learned for more than three years. Their hearts were shattered, their spirits broken. In such a state, how could they have mustered the courage and resolve to approach the tomb, confronted an armed guard, and carried out a bold theft?

While Jesus was being interrogated, Peter was in the courtyard outside Caiaphas' house. The uproar had drawn crowds of onlookers and agitators, shouting for punishment. When a servant woman confronted Peter, saying he had been with Jesus, he denied it: "He denied it before them all, saying, 'I do not know what you are talking about.'" (Matthew 26:70)

If Peter, one of Jesus' closest companions, was afraid of a maidservant, how can we believe that he later summoned the courage to confront the Roman guard—men known for their brutality and efficiency?

These soldiers did not flinch as they drove nails into human flesh. They had no hesitation placing a crown of thorns on a man who had just been scourged. They obeyed without question when ordered to lash a defenseless prisoner with the *flagrum taxillatum*[155], a whip designed to tear skin and muscle—stopping only when death was near.

---

[155]The instrument used for flagellation was the *flagrum taxillatum*, which consisted of a short wooden handle to which were attached three leather straps, each approximately twenty inches long. At the ends of each strap were two elongated lead balls, tightly bound together. In some cases, instead of lead balls, the straps bore *talli*—the astragalus (ankle bone) of a ram. However, the version with lead balls was the most used.

According to Hebrew law, the maximum number of lashes allowed was forty. To avoid accidentally exceeding this limit, the Jews would typically administer only thirty-nine lashes. However, Jesus was flogged by Roman soldiers under Roman authority. Therefore, he was scourged more roman—according to Roman custom—which placed

These were the very men who stood between the disciples and the tomb. Were they overcome by a group of untrained, frightened men?

The record tells us otherwise. The guards survived. On the day of the resurrection, they were unharmed, and instead of defending their actions, they went to the chief priests: "Some of the guards went into the city and reported to the chief priests all that had happened." (Matthew 28:11)

If there had been a confrontation, if the disciples had fought and defeated them, the guards would not have gone to the religious authorities seeking cover. They would have been summoned for punishment or interrogated by Roman officials. Instead, they accepted a bribe, and a story was crafted to cover their failure.

TENTH THESIS: FROM HONEST TO VANDALS?

The disciples were the people closest to Jesus throughout His apostolic ministry. They listened to His teachings directly and were gradually transformed in character and temperament, being formed to become builders of the Kingdom of Heaven. Over time, they moved away from their traditional Jewish worldview and embraced the one the Master taught them—by His words and His example.

Where, then, would they have found the spirit of vandals, deceivers, or criminals, capable of executing such a Machiavellian plan as stealing the body of their Master—without any respect for the dead or for His family—just to deceive the public with a false resurrection?

Is it possible to believe that they intended to build a Church on the vilest of lies? Would Mary, the mother of Jesus, have gone along with a plan that began with desecrating her Son's tomb and then proclaiming

---

no legal limit on the number of lashes. The only requirement was that the condemned remain alive.

There were two reasons for this: first, so that Jesus could be presented to the public in a pitiable state, in hopes of eliciting sympathy (as was Pilate's intention); and second, in the event of a death sentence, to ensure that he would survive the journey to the execution site and be crucified alive, as Roman law required.

to the world that He had risen? Let us remember that Mary remained part of the early Church. She was present with the apostles at Pentecost:

> Then they returned to Jerusalem from the mount called Olivet, near Jerusalem, a sabbath day's journey away. When they entered the city, they went to the upper room where they were staying: Peter and John and James and Andrew... All of these with one accord devoted themselves to prayer, together with some women, including Mary, the mother of Jesus. (Acts 1:12–14, see also Acts 2:1–4)

According to Jewish law, a human corpse was the greatest source of ritual impurity:

> Anyone who touches the corpse of any human being will be unclean for seven days. Such a person must purify himself with the water of purification on the third and on the seventh day, and then he will be clean. But if he does not purify himself on the third and seventh day, he shall not be clean. Anyone who touches the corpse of a human being who has died and does not purify himself defiles the tabernacle of the Lord. That person shall be cut off from Israel, because the water of purification was not sprinkled upon him. (Numbers 19:11–13, 16)

There is no indication, either in the Bible, in apocryphal writings, or in secular literature, that the apostles ever ceased to be obedient to the Law of Moses—just as Jesus had been.

So, what evidence exists to suggest that, within seventy-two hours, the apostles went from honoring the Law to flagrantly violating it, disregarding the instructions found in Numbers 19:11–16?

We know that, on the day of the first appearance of the risen Lord, the disciples were all gathered in one room. If they had become ritually impure by touching the body of Jesus, why would they have gathered, mingling the unclean with the clean?

The three years they had spent with the Lord had been like a spiritual roller coaster—marked by moments of great joy, fear, confusion, and reflection. His teachings were not always easy to understand, and it took time for them to fully grasp His message.

Yet there was one teaching they consistently struggled to accept: the announcement of His passion, death, and resurrection.

If Jesus truly was the Messiah, the Son of God, one with the Father, how could He be judged, condemned, and crucified—and then rise again? How could God be killed?

The disciples' relationship with Jesus evolved in stages. At first, He was simply the one pointed out by John the Baptist. Then, after witnessing many signs, they came to regard Him as a prophet. Over time—and with difficulty—they came to recognize Him as the Messiah.

But on that tragic Friday, they lost that conviction. In their eyes, He ceased to be the Messiah and reverted to being a prophet, as shown in the words of the disciples on the road to Emmaus:

> Then one of them, named Cleopas, answered him, 'Are you the only visitor in Jerusalem who does not know the things that have taken place there in these days?' He asked, 'What things?' They replied, 'The things about Jesus of Nazareth, who was a prophet mighty in word and deed before God and all the people.' (Luke 24:18–19)

Upon His death, the disciples no longer believed He was who He had claimed to be. They gave Him a burial appropriate for a man, not for the divine. They believed they had seen the last of Him. The story, as far as they knew, had ended at the cross.

If they no longer believed He was the Messiah, what motivation could they have had to stage a resurrection? Why would they risk the legal, religious, and military consequences of stealing what was the best-guarded body in history?

Even the beloved disciple admits that, upon seeing the empty tomb, it was only then that he began to understand: "They did not yet understand the Scripture which stated that He had to rise from the dead." (John 20:9)

And it was Jesus Himself, on the very day of His resurrection, who had to explain the Scriptures to them before they understood: "Then they said to each other, 'Were not our hearts burning within us while He was talking to us on the road and explaining the Scriptures to us?'" (Luke 24:32)

Why would the disciples go to such extreme lengths—risking defilement, arrest, and punishment—to orchestrate a false resurrection that they themselves did not yet believe in?

Like Martha, the sister of Lazarus, they believed in a general resurrection at the end of time, but not in the immediate resurrection of Jesus:

> Jesus said to her, 'Your brother will rise again.' Martha said to him, 'I know that he will rise again at the resurrection on the last day.' (John 11:23–24)

ELEVENTH THESIS: WHERE ARE HIS REMAINS?

Jesus gained popularity among the Jewish people not so much for the message of good news that He preached, but for the miracles He performed. He healed, restored, and fed them. As He Himself said: "Amen, amen, I say to you, you are looking for me not because you have seen signs but because you ate your fill of the loaves." (John 6:26)

The people recognized that a special power flowed from Him—something that transformed everything it touched:

> When she had heard about Jesus, she came up behind him in the crowd and touched his cloak. For she said, 'If I can only touch his clothing, I shall be healed.' Immediately, the flow of blood stopped, and she felt in her body that she was healed of her affliction. Jesus, aware at once that power had gone out from him, turned around in the crowd and asked, 'Who touched my clothing?' (Mark 5:27–30)

The thought of this woman perfectly mirrors the mindset of the crowd. This is why they sought Him and followed Him from place to place. In most Gospel scenes, Jesus appears surrounded by multitudes—people who loved to hear Him and see Him challenge the religious elite. Yet always present was the hope of a miracle, a healing, or a solution to their most urgent needs.

Many lived in expectation of His next arrival in their region, hoping merely to touch Him: "All the crowd sought to touch him, because power came forth from him and healed them all." (Luke 6:19).

The Jews of His time venerated the tombs of prophets and other holy men, especially righteous martyrs: "Woe to you, scribes and Pharisees, hypocrites! You build the tombs of the prophets and adorn the memorials of the righteous." (Matthew 23:29), "My brothers, I can tell you confidently that the patriarch David died and was buried, and his tomb is here among us to this day." (Acts 2:29) or "Then Simon built a tomb for his father and his brothers and made it high so that it could be seen. He constructed seven pyramids facing one another for his father, his mother, and his four brothers." (1 Maccabees 13:27–28)

Why, then, is there no historical, Christian, or secular record of a site where the body of Jesus was venerated?

When Jesus entered Jerusalem on the Sunday before His Passion, riding a donkey, the crowds rejoiced, laid down palms, and cried out "Hosanna". His arrival drew great attention:

> When He entered Jerusalem, the whole city was stirred and asked, 'Who is this?' And the crowds replied, 'This is the prophet Jesus from Nazareth in Galilee.' (Matthew 21:10–11)

This was the reality of Jesus' reception: the people acknowledged Him as a prophet, but not as the Messiah. Yet among all the prophets of Israel, none performed as many works and miracles as Jesus. And still, while the tombs of those prophets were venerated, the tomb of Jesus was not.

Why?

The answer is simple: because the tomb of the Master was empty after three days, and it was never used again.

We know the final resting places of Abraham, Mohammed, Buddha, Confucius, Lao-Tzu, and Zoroaster. But where are the remains of Jesus?

Could this not be further evidence that His friends did not possess the body?

TWELFTH THESIS: THE SHEET IN THE GRAVE

Most biblical scholars identify the apostle John—one of the two sons of Zebedee and Salome, the younger brother of James, and companion of Simon Peter—as the beloved disciple mentioned in the Gospel that bears his name. He was among the first called by the Master to be one of the Twelve. This was unusual for the time, as it was customary for disciples to choose their teachers, not the other way around: "You did not choose me; no, I chose you." (John 15:16)

John is believed to have been the youngest of the apostles. For this reason, Jesus developed a special affection for him—thus the nickname "the beloved disciple."

John was present with the Master during some of His most intimate and significant moments: In the house of Jairus, the synagogue leader, when Jesus raised his daughter from the dead, or, on Mount Tabor, during the Transfiguration, or, In the Garden of Gethsemane, where Jesus prayed in agony before His Passion.

John, along with Peter, was chosen to prepare the Last Supper, and during the meal, Jesus invited him to sit at His right. At the foot of the cross, Jesus entrusted to John the care of His mother. After the resurrection, John was also a witness to one of the appearances of the Lord, as well as the miraculous catch of fish at the Sea of Tiberias.

This disciple had seen the Master: Raise the dead, walk on water, feed thousands with just a few loaves and fish, heal every kind of illness, give sight to the blind, restore speech to the mute, enable the paralyzed to walk.

John had spent three years at Jesus' side—day and night—sharing countless conversations, many of which are recorded in Scripture, and many more that are not, but certainly took place.

And yet, despite all of this, John still did not believe that Jesus was the Son of God—at least not fully.

The Gospels describe the turning point for several disciples—those moments of revelation when they came to believe in the resurrection, and thus in Jesus as the Messiah:

- For Thomas, it was when Jesus invited him to touch the wounds in His hands and side.
- For the two disciples on the road to Emmaus, it was when He broke the bread during their shared meal.
- For others, it was when Jesus appeared to them in the upper room, where they had hidden in fear.

So then—what was the breaking point for John, the disciple Jesus loved in a unique and personal way?

Together with Peter, John was one of the first disciples to visit the tomb after Mary Magdalene announced that the Lord had risen. Upon entering, they noticed that "everything" was in its place—everything except Jesus.

> The two of them ran together, but the other disciple ran faster than Peter and arrived at the tomb first. He bent down and saw the linen clothes lying there but did not go in. Then Simon Peter arrived after him and entered the tomb. He saw the linen cloths lying there and the cloth that had covered Jesus' head, not lying with the other cloths but rolled up in a separate place. (John 20:4–7[156])

The evangelists use the words "cloth" or "sheet" to refer to the linen shroud that Joseph of Arimathea had purchased to wrap the body of Jesus. In the account given by the beloved disciple, a particular detail is emphasized—the word "lying" (or "stretched out") is repeated, underscoring its importance.

Upon entering the tomb, the disciples were astonished to see that the body was gone—but the shroud remained in place. It was lying where the body had been, seemingly untouched, but collapsed in on itself, as though the body had passed through it. It appeared "deflated." This

---

[156]Most Bible translations, unfortunately, lead readers to imagine Jesus wrapped like an Egyptian mummy, which obscures the true reason the beloved disciple believed. In 2010, the Spanish Episcopal Conference introduced a new Spanish translation of the Bible, now used as the official text of Sacred Scripture proclaimed in the liturgy. In this version, the account of the cloths found in the tomb has been improved through a more accurate rendering of the original Greek texts.

remarkable detail is what prompted John to record the reaction: "He saw and believed." (John 20:8)

This disciple—who had been especially close to the Lord, who had accompanied Him during His most intimate moments—finally believed when he saw the cloth "lying there." What he saw was not a disorderly pile, but the contours of the linen cloth still shaped around the form it once held: the nose, cheekbones, chin, torso, and limbs, all visible in soft relief. It lay exactly where the body had been placed on Friday evening.

For John, this crucial detail not only disproved the rumor that the body had been stolen—after all, what thief would have carefully arranged the burial cloths? —but pointed directly to the miracle of the resurrection.

The Gospel of John is also the only one to mention, alongside the linen cloth, the "shroud." The Greek word used here refers to a "sweat cloth" or "handkerchief"—a piece of fabric smaller than the linen sheet, larger than a typical handkerchief, but smaller than a towel. It was a common item in men's attire during Jesus' time and was used, particularly in burial customs, to wrap the face, primarily to keep the jaw from falling open. It was the first element used in the burial preparation.

This same Greek word appears in: John 11:44, in the resurrection of Lazarus: "The dead man came out, his hands and feet bound with strips of cloth and his face wrapped in a shroud.". Luke 19:20, in the parable of the talents[157], where a servant keeps his coin wrapped in a cloth. Acts 19:11–12, where Paul's garments were used for healing: "Even handkerchiefs or aprons that had touched his body were taken to the sick, and their illnesses were cured, and evil spirits departed from them."

In John 20:7, particular emphasis is placed on the position of this shroud: unlike the linen sheet, it was not lying flat but had been rolled up and placed separately.

What the evangelist is describing is this: the body of Jesus passed through both the shroud and the linen cloth. Then He removed the

---

[157]The third servant returns the talent to his master, saying: "Here is your talent; I kept it hidden in a handkerchief [or shroud]."

shroud, folded it, and set it aside, in a different place from where His body had been laid.

This was the moment of revelation for John. This is what led him to believe in the resurrection, to recognize that Jesus was indeed the Messiah, the Son of God.

"He saw and believed." (John 20:8)

THIRTEENTH THESIS: MARTYRS

At the beginning of the 20th century, after the fall of the Russian Tsar, a progressive ban on the practice of any religious rites was imposed across the country. Still, certain traditions continued in secret. One of these was the baking of Easter bread, known as *Kulich*, considered the festival of all festivals by the faithful.

Despite increasing restrictions, believers continued to bake *Kulich* in varied forms and styles, sharing it in quiet, home-based celebrations with family and friends. However, as communism deepened its grip on every corner of the country, authorities began to aggressively suppress all religious expressions. In some areas, even baking *Kulich* was prohibited.

One such case occurred in a small village near Kyiv, where in the early 1930s, the local authorities ordered the confiscation of all flour and the shutdown of ovens. Yet, before the raids began, some villagers managed to hide enough flour and ingredients to prepare the sacred bread for the upcoming Easter.

One of the most powerful men in the world at that time was Nikolai Ivanovich Bukharin[158], a Russian communist leader and participant in

---

[158]Nikolai Ivanovich Bukharin (Moscow, October 9, 1888 – March 15, 1938) was a Russian Marxist revolutionary, politician, economist, and philosopher. A prominent figure in the Bolshevik leadership, he served on the Politburo until 1929 and was editor of Pravda. Throughout the 1920s, he was recognized as the chief theoretician of Soviet communism and led the Comintern from 1926 to 1929. Between 1925 and 1928, Bukharin was one of the leading Soviet figures alongside Joseph Stalin. He was a principal advocate for gradual economic modernization and the transition to socialism. During 1928–1929, he emerged as the most prominent representative of the so-called "right-wing opposition" within the Communist Party.

the Bolshevik Revolution. On Easter Day, 1930, Bukharin addressed a mass gathering of workers in a nearby town near Kyiv to promote atheism.

For two hours, he launched an aggressive speech using the "heavy artillery" of anti-religious propaganda, hurling insults, and alleged evidence against the existence of Jesus and the truth of His legacy. When he finally concluded, he stood confidently, believing that all that remained was a pile of broken faith among the listeners.

He asked if anyone had anything to say.

A local priest, hoping to encourage the faithful and inform them that, despite the ban, the *Kulich* had been baked in secret, asked for a moment to speak. He was given three minutes. He responded that he would need far less than that.

He looked out over the crowd and, with a resounding voice, proclaimed the traditional greeting of the Orthodox Church: "Jesus Christ is risen!"

In a powerful, unified response, the crowd rose to its feet and answered like thunder: "He is risen indeed!"

All the evidence found in the New Testament and in early Church literature confirms that the central message of the Gospel was not "Follow the Master's teachings and live virtuously," but rather: "Jesus Christ rose from the dead."

This is what the apostles went out to proclaim—and it cost them their lives.

What greater testimony could there be that the humble carpenter of Nazareth was neither mad nor a deceiver when He said: "The Father and I are one"? (John 10:30)

Although the Bible records the deaths of only two disciples—Judas Iscariot, the betrayer who hanged himself (cf. Matthew 27:5), and James

The Greater[159], who was beheaded by order of King Herod (cf. Acts 12:2)—Tradition tells us that all the others were martyred.

While the places and circumstances of their deaths vary across sources, one consistent point emerges: they all died as martyrs.

Below are the accounts of their deaths, according to the most widely accepted traditions.

John, the beloved disciple of the Lord and brother of James *The Greater*, is traditionally recognized as the author of the Gospel that bears his name, the Book of Revelation (Apocalypse), and two epistles. According to early Christian tradition, he survived being plunged into a vat of boiling oil, a punishment ordered by Emperor Domitian for preaching the Gospel. When this attempt on his life failed, the emperor sentenced him to forced labor in the mines on the island of Patmos. After some time, John was released, and he later died peacefully on the island of Ephesus.

The martyrdom of Peter was foretold by Jesus Himself, and the evangelist John recorded it in allegorical language: "Jesus said this to indicate the kind of death by which Peter would glorify God." (John 21:19)

Peter died in Rome, crucified upside down by order of Prefect Agrippa, an official under Emperor Nero. According to tradition, Peter requested this form of crucifixion because he felt unworthy to die in the same manner as his Master.

Andrew, Peter's brother and son of Jonah, was martyred in Achaia, Greece, in the town of Patras. When Governor Aepeas' wife and brother converted to Christianity, the governor became enraged. He arrested Andrew and sentenced him to death by crucifixion. Out of humility, Andrew requested a different form of cross than that of Jesus, and so he was crucified on an X-shaped cross, which to this day is known as the Cross of Saint Andrew, one of his traditional symbols. His martyrdom is dated to November 30 in AD 63, under Nero's reign.

---

[159]Known as "the Greater," he was the brother of the apostle John—both sons of Zebedee and Salome. In some Bible translations, his name appears as James.

James *The Less* (also known as Jacobus), the half-brother of Jude Thaddeus and son of Alphaeus and Mary, was martyred in the year AD 62. The high priest Annas II ordered him to publicly deny Jesus, but instead James began preaching the Gospel from the top of the temple. Enraged, the Pharisees and scribes pushed him from the heights. Since the fall did not kill him, they began to stone him as he prayed on his knees for his executioners. Finally, he was killed by a blow to the head with a mace.

Jude Thaddeus, also known as Lebbaeus, son of Cleophas and Mary, was beheaded with an axe in the city of Suamir, in Persia.

Matthew, also called Levi, the son of Alphaeus and author of one of the four Gospels, was martyred in Nadaba, Ethiopia. He had opposed the marriage of King Hirciacus with his niece Iphigenia, a Christian convert. For this, he was beheaded at the conclusion of a sermon around the year AD 60.

Simon *The Canaanite*, also known as the Zealot, was martyred in Suamir, Persia, where he was sawn in half.

Philip, originally from Bethsaida, preached throughout Asia and later in Heliopolis, Phrygia (present-day Turkey). He was imprisoned and later crucified there in the year AD 54.

Bartholomew, also called Nathanael, son of Talmai, was martyred in Albana, Armenia. He was first crucified, then taken down before death, flayed alive, and finally beheaded. Because of this, ancient Christian art often depicts him carrying his skin like a cloak, draped over his arms.

Thomas, known as Didymus and sometimes referred to as the doubter, was martyred on the Coromandel Coast of India. His body was found pierced with spears, indicating a violent death for his faith.

The term *kamikaze*, of Japanese origin, was first used by American translators to describe the suicidal attacks carried out by pilots of the Imperial Japanese Navy against enemy ships toward the end of World War II. These attacks were intended to stop the advance of the Allied forces across the Pacific and prevent them from reaching Japanese shores. To that end, aircraft loaded with 250-kilogram bombs were deliberately crashed into their targets to sink or severely disable the vessels.

This term has also been used by some journalists to describe certain jihadist terrorists, who seek to kill as many "infidels" as possible while being fully aware that their actions will result in their own deaths. These individuals are motivated by the belief that Allah will reward them in paradise with seventy-two virgins (houris), rivers of wine, honey, and milk, winged horses made of gold and rubies, and other delights meant to satisfy them for eternity.

Beginning in 2009, more than twenty Tibetan monks chose to immolate themselves in protest of the Chinese government's ban on the return of the Dalai Lama to his native Tibet. For these monks, self-immolation became the only way to draw global attention and pressure the occupying forces to withdraw from their homeland.

In all three cases—jihadist terrorists, Japanese kamikaze pilots, and Tibetan monks—the individuals involved engaged in acts that, while technically suicidal, are often not classified as suicide by those who perform or endorse them. Their religious traditions condemn suicide, but these acts are viewed not as personal despair, but as a sacrifice for a greater, collective cause. In such contexts, the rules are reinterpreted.

However, the case of the apostles is entirely different.

When a jihadist leaves home wearing a bomb vest, fully aware that he will die with his victims, or when a kamikaze pilot intentionally crashes his plane into an enemy ship, or a Tibetan monk binds himself in barbed wire to prevent rescue during his self-immolation, each of them knows with absolute certainty that death is imminent. These are deliberate acts with death as the goal—technically and ethically, suicides.

This was not the case with the apostles.

When they proclaimed the resurrection of the Lord, they were fully aware that their message would bring them trouble, and even death—as it did. But they did not seek death, nor did they wish for it. They simply could not deny what had become impossible for them to deny: that they had seen the Master with their own eyes after He had been buried in the tomb so generously offered by Joseph of Arimathea.

The apostles did not give their lives to defend a doctrine, or to preserve the teachings of Jesus, or to protect a nascent Church, much less to safeguard a religion.

They were driven by the reality of the resurrection. They went out into the world to proclaim what they had witnessed firsthand—to recount the life-changing moments they had shared since Jesus of Nazareth entered their lives and called them to follow Him in the most extraordinary experience of their time.

They gave testimony, told the truth of their experience, and for this— they were martyred.

CONCLUSION

The Book of the *Acts of the Apostles*, written by Luke—the same author of the Gospel that bears his name—narrates the foundation of the Catholic Church and the spread of Christianity throughout the Roman Empire. After the apostles received the Holy Spirit on Pentecost, they began organizing daily gatherings in homes to celebrate the commemoration of the Lord's Last Supper: "Day after day they met in the temple, and in their homes they broke bread and shared their food with glad and generous hearts." (Acts 2:46)

The word "last" before the word supper should naturally evoke a sense of finality—a solemn farewell. And indeed, in the case of the Lord's Supper, it marked the beginning of the end, as the events that led to His death followed shortly thereafter.

Yet why did the early Christians not gather in mourning, grief, or lamentation to commemorate that moment? Why did they instead celebrate it with joy?

If there had been no resurrection, there would have been nothing joyful to celebrate.

But Jesus had foretold this very transformation:

> You are confused because I said: 'A little while and you will not see me, and again a little while and you will see me.' Amen, amen, I say to you, you will weep and mourn while the world rejoices. You will grieve, but your grief will be turned into joy. (John 16:19–20)

For many Catholics, the resurrection is just one more article of faith—something believed more out of habit than conviction. In their hearts, they may wonder: How can it be proven that Jesus Christ rose from the dead if it happened so long ago? Or how can we trust that the apostles did not write only what served their purpose?

When these questions arise, many prefer to avoid them, fearing that they challenge the apostles' honesty. Yet the evidence presented in this chapter provides a foundation strong enough to dispel doubt.

As stated earlier, the Bible is not the only source that testifies Christ was crucified, died, and was buried, and that on the third day, many witnesses reported seeing Him alive, and some of them interacted with Him.

The Gospels give us abundant detail—revealing the honesty, spontaneity, and even the naive of the writers. But they are not the only confirmation of these events. Our faith in the resurrection of the Lord is no longer a leap into the void, but rather a journey along solid ground, supported by strong evidence.

Why, then, does Paul say that if Christ did not rise, our faith is in vain?

Because without the resurrection, there would be no Christianity. There would be no apostles, no Church, and no hope of eternal life. We would still be waiting anxiously for the one who would redeem us from sin and open the way to the Father's house.

The resurrection of the Lord is decisive.

To understand this, we go back to Abraham, the first man in history to whom God revealed Himself. Before him, people believed in many gods—deities made of stone, metal, or natural forces. But when God spoke, Abraham listened. And the Lord made a promise:

> Leave your country, your family, and your father's house, and go to the land that I will show you. I will make you a great nation. I will bless you and make your name great, so that you will be a blessing. I will bless those who bless you, and I will curse those who curse you. Through you all the families of the earth shall be blessed. (Genesis 12:1–3)

This was the foundation of the promise to the people who would become known as Israel. Although the blessing was intended to reach all the families of the world, Abraham's descendants would become a "great nation." God asked only one thing in return: faithfulness.

Every time the Israelites remembered this promise—especially the promise that they would become a great nation—they imagined it in the image of the military and economic power of their age: the Egyptians, the Babylonians, the Greeks, or the Syrians, depending on the period.

Yet despite their expectations, God remained faithful to His covenant, while the Israelite people often did not. And so, generation after generation, they continued to long for the day when they would become great.

Historically, a series of prophets foretold the arrival of a man who would restore dignity to the people of Israel, bring good news to the poor, proclaim liberty to captives, give sight to the blind, and set the oppressed free. This man would not be just any man: He would be God made flesh, the one we would call Emmanuel—the Messiah.

As explained in chapter two, there were hundreds of signs—prophecies given by the prophets—that would help the people identify the long-awaited Messiah. It was also shown that all these predictions were fulfilled in the person of Jesus.

One might assume this would have been sufficient for the people to recognize Him and rejoice, knowing that God was now among men. But the spiritual blindness was such that they did not recognize Him. It fell to Jesus Himself to reveal that He was the one they had been waiting for.

How did the most educated and religious class of Israel—the very ones who knew the Law and the Prophets by heart—respond to this claim?

They regarded Jesus as a madman, an impostor, a blasphemer.

The Jews expected the Messiah to be, at the very least, a figure like King David—a name that in Hebrew means "the beloved" or "the chosen one of God." David, born in Bethlehem (the same city where Jesus was born) in BC 1040, and who died in Jerusalem in BC 966, was the son of Jesse and Nitzevet. As the youngest of seven sons, he was destined for the low work of shepherding sheep. Yet, he went down in history as a just, brave, and passionate king—a warrior, musician, poet, described in sacred Scripture as "blond, with beautiful eyes, prudent, and of good appearance." Like all great men, he was not without sin, but he unified the twelve tribes of Israel, completing what his predecessor Saul had begun. Still, during the reign of David's grandson Rehoboam, the kingdom fractured once more.

This was the resume that the cultured elite of Israel expected in their Messiah.

A poor carpenter, with no wealth and no army, could hardly be imagined as the fulfillment of that hope.

However, the many and extraordinary miracles performed by Jesus caused both confusion and intrigue among the Sanhedrin. They saw Him restore sight to the blind, speech to the mute, hearing to the deaf, strength to the paralyzed, and life to the dead. Clearly, He was not an ordinary man—His works went beyond the natural.

But if His miracles intrigued them, His words infuriated them.

Jesus' relationship with the religious authorities of Israel fluctuated between intrigue and outrage. At times, they simply ignored Him. But whenever an encounter became inevitable—especially during His visits to the Temple in Jerusalem—Jesus was uncompromising. He openly rebuked them for killing the spirit of the Law given through the prophets. He accused them of turning the Law into a burden, one they themselves were unwilling to bear. He called them: "Hypocrites," "evildoers," "faithless," "fools," "brood of vipers," and "blind guides." He even compared them to whitewashed tombs—beautiful on the outside, but full of corruption within.

One day, the Pharisees and the scribes decided to challenge Him. They demanded another sign, another miracle to prove that He truly was the Messiah. Jesus responded:

> An evil and unfaithful generation seeks a sign, but no sign will be given it except the sign of the prophet Jonah. For just as Jonah was in the belly of the great fish for three days and three nights, so will the Son of Man be in the heart of the earth for three days and three nights. (Matthew 12:39–40)

The Master Himself told them that the only sign He would give was His resurrection, not His miracles.

If He rose from the dead, it meant He was neither mad nor deceitful, but truly God incarnate. It meant that everything He had said was the purest truth. It meant that He would not have continually quoted the Scriptures unless those writings were indeed the words of God, entrusted to the prophets by the Father.

It meant that the Law had been reborn, animated now by a new spirit.

It meant that the waiting was over—the one who would redeem us from sin had already come.

It meant that the hope of eternal life with the Father was now a reality.

It meant that the Church, foretold as a bridge between earth and heaven, had been born.

It meant that we could now place our full confidence in everything He promised and hold fast to it.

It also meant that we could call Jesus our brother, Mary our mother, and God our Father.

This is why Paul wrote: "If Christ has not been raised, then our preaching is in vain, and your faith is also in vain." (1 Corinthians 15:14)

But He did rise.

In the two thousand years since the resurrection of Christ, countless theories have been proposed to distort or reinterpret that event. Some claim that it was merely a story invented by a group of disciples determined to start a new religion based on Judaism—regardless of the cost.

Yet such claims ignore the vast body of evidence—from both Christian and non-Christian sources.

The tomb of Jesus had the imperial seal of the highest Roman authority, making it illegal for anyone to tamper with it without authorization. It was also guarded by highly trained Roman soldiers, operating under the strictest codes of military conduct, watching day and night over the only access to the tomb.

And yet, three days later, those same soldiers went to the chief priests seeking help to create an alibi—to avoid punishment for allowing the body to vanish from the tomb.

Of course, we cannot say that resurrection is the only possible explanation when a body disappears from its resting place. No—such a conclusion should never be entertained lightly.

But this case is unique.

The idea of resurrection could not even be considered unless the event had been clearly prophesied, and unless the deceased had claimed to be God, possessing the power and authority to overcome death and rise again by His own will.

In this chapter, I have presented thirteen carefully developed theses, each one coherent and consistent with the accumulated facts preserved in historical literature, logical reasoning, and Sacred Scripture.

In the second chapter, I demonstrated that the Holy Spirit is the author of the Bible, which means that this unique and sacred book cannot be excluded when gathering evidence to understand the mystery of the empty tomb.

I also cited the testimonies of historians such as Flavius Josephus, Cornelius Tacitus, and Gaius Plinius Caecilius Secundus (Pliny the Younger)—whose writings have survived through the centuries. Though they do not provide the level of detail found in Christian testimony, they nevertheless confirm the essentials—the heart of the matter:

- That Christ was crucified by order of Pontius Pilate
- That He was buried outside the city of Jerusalem, near the place of His execution.
- And that, days later, many people claimed to have seen Him alive.

Likewise, in the previous chapter, I demonstrated that the prophecies spoken by the prophets over several centuries—meant to help identify the Messiah—were fulfilled with the coming of Jesus. I also showed how it is mathematically impossible that these prophecies could have been fulfilled by chance at Jesus' birth if He were not the Messiah.

I analyzed the facts from every plausible angle: from the possibility that the women went to the wrong tomb, to the suggestion that Jesus had not actually died, to the theory that the body had been stolen. I presented several hypotheses proposed by anti-Christian groups and individuals regarding the empty tomb. Yet, when these theories are confronted with the full body of evidence, each one falls apart.

In every case, there is at least one key fact that does not "add up":

- That He did not die?
- That people saw a double of Jesus?
- Then where is His corpse?
- Why did the guards have to seek an alibi?

If the body was stolen, someone had to have done it. I examined the only two sides that could have carried this out—His friends or His enemies. But neither theory fully aligns with the facts.

Against all logic and reason, the resurrection remains the only explanation that fully satisfies the evidence.

Jesus made the boldest claim in history: He said He was God.

Not that He was King David, Isaiah, Moses, or Abraham—but God Himself.

Unsurprisingly, many considered Him insane. But after witnessing His many miracles, the people asked Him for one final, conclusive sign— something that would leave no doubt that He was who He claimed to be. And He told them: the resurrection would be that sign. He delivered on that promise. Jesus proved He was God. He proved He was the Messiah foretold by the prophets.

The voice of God, spoken through those holy men, is recorded in the Sacred Scriptures, as is His own voice, through His Son, Jesus Christ.

Can we trust that communication?

Without a doubt.

EPILOGUE

Then Paul stood before them in the Areopagus and said: "Men of Athens, I have seen how religious you are. For as I walked around, looking carefully at your shrines, I noticed among them an altar with the inscription, 'To an Unknown God.' What, therefore, you worship as unknown, I now proclaim to you. "The God who made the world and everything in it, the Lord of heaven and earth, does not dwell in shrines made by human hands. Nor is He served by human hands as though He were in need of anything. Rather, it is He who gives to everyone life and breath and all other things. From one ancestor, He created all peoples to occupy the entire earth, and He decreed their appointed times and the boundaries of where they would live. "He did all this so that people might seek God in the hope that by groping for him they might find him, even though indeed He is not far from any one of us. For 'In him we live and move and have our being.' As even your own poets have said, 'We are all his offspring.' "Since we are God's offspring, we ought not to think that the deity is like an image of gold or silver or stone, fashioned by human art and imagination. God has overlooked the times of human ignorance, but now He commands people everywhere to repent, because He has fixed a day on which He will judge the world with justice by a man whom He has appointed. He has given public confirmation of this to all by raising him from the dead."

ACTS 17:22-31

After reading this book, I believe we can agree on the great importance of the three central questions I chose to answer—questions whose answers deeply affect our lives: That God exists, that He is the Creator of all things visible and invisible, that He has spoken through the prophets, that He became man, that He taught us what love is, that He died on the cross and rose again on the third day, that He instructed us to call God our Father, Mary our Mother, and Jesus our Brother—in short, that we belong to a heavenly family.

Since the moment He created us, God has maintained ongoing communication with humanity. Among the many ways He has spoken to us, the Holy Scriptures hold a privileged place.

Throughout this book, the existence of God and His two most fundamental roles—as Creator and Father—have become evident. But in honest and deep reflection, we must ask: What should the certainty of God's existence as Creator mean for us?

In the first chapter, I explained that when Charles Darwin introduced his theory of evolution, one of its side effects was to dethrone humanity from its special place in creation. Until then, we had believed ourselves to occupy a privileged position as the only species created in the image and likeness of God. Darwin's theory challenged this by presenting us as just one more species, slightly more fortunate than others.

In contrast, I have provided abundant and converging evidence for the existence of a Creator—a Creator who, from the beginning, had in mind all of creation, and within it, humanity as His greatest work, as revealed by the Scriptures. We can once again claim our place of honor, with crown and scepter, not as the result of chance or natural accident, but as the masterwork of a divine plan, far beyond our comprehension.

Every creation has a purpose, and the universe is no exception. God created us with a definite intention. But how can we discover it?

Many people believe that life's purpose involves achieving happiness, self-realization, personal success, travel, wealth, or legacy. But being successful and fulfilling life's purpose are not the same thing.

To illustrate this, consider the story of Hugh S. Moorhead, a philosophy student at the University of Chicago, who, before graduating, wrote to 250 of the most influential thinkers of his time—philosophers, scientists, writers, and intellectuals—asking a single question:

What is the purpose of life?

Some replied as best as they could. Others admitted they had never considered the question until he asked. And some answered with complete honesty: they had no idea. A few even requested that he please let them know if he ever found the answer.

He later compiled these responses in his book *The Meaning of Life* (1988).

This simple yet profound question is not easy to answer.

Imagine I give you a strange metallic device you have never seen. It is heavy, cube-shaped, and you can wrap your hands around it. You decide to use it as a paperweight, thinking that it must be its purpose. But then, you ask the inventor, and he tells you it is a 3D image projector. You must place your finger in a corner for five seconds, and then it opens and displays breathtaking three-dimensional images.

You would immediately realize: What a waste!

You were using something so advanced for something so basic, simply because you did not understand its purpose.

Now, consider this:

When a couple decides to have a child, they are not thinking of creating the next president, or a brilliant scientist, or the next pope. They are thinking of themselves—of their shared love and commitment. They imagine their child's future with dreams and hope, knowing that eventually the child will make his own decisions. And they trust that, with enough love, they will have done their part well.

Similarly, we were created by God, not for our own autonomous purposes, but for His. As Bertrand Russell, the philosopher, mathematician, and atheist writer, once said: "Unless you assume a God, the question of life's purpose is meaningless."

We will not discover the purpose of life by looking within ourselves, as many self-help and motivational books suggest. We did not create ourselves, so we cannot find our reason for being solely by introspection.

We were created by God and for God, just as our parents created us by them and for them. As part of His divine plan, the purpose of our life is to let Him use us for His purposes, not for us to use Him for ours.

Now, if we reflect again—deeply and sincerely—on the certainty of God's existence, not just as Creator but as Father, what should that mean for us?

Every relationship is built on communication. With God, it is no different.

We speak to Him through prayer, and He speaks to us through His Word.

I have shown that the Bible is a living book, inspired by the Holy Spirit, and that it contains the words every good father desire to share with his children—for their well-being and guidance.

Let us return to the example of the mysterious metal box. Without the instruction manual, you used it as a paperweight. But once you knew its true function, you were astounded by what it could do. In the same way, the Bible is our manual—our guidebook for life. Yes, it has some passages that are difficult to understand due to the historical, cultural, and linguistic distance. But if those difficulties are what make you reluctant to read it, I can summarize its message in a single word: love.

In three words: God is love.

Or, if you prefer something more complete: let us love one another as Jesus loved us.

And how did He love us?

The closest human comparison I can think of—although I know it falls short—is the love of a mother for her child. It is not uncommon to hear a mother say that she would gladly give her heart for her child: "Just tell me the time and place, and I'll be there."

She would not just show up—she would be happy to do so.

That is the kind of love that Jesus gave us.

When we speak about "love", we encounter two distinct challenges.

The first is the overuse of the word. Its meaning has been diluted because we use it too easily and in too many contexts. We speak of love for our country, our work, a piece of art, a pet, a meal, or even a favorite restaurant. This widespread use has gradually eroded the depth of the word's original intent.

The second issue is linguistic limitations. For example, in Spanish, we have two words to express affection at different intensities: *amar* and *querer*. However, when Spanish is translated into English, both are usually rendered simply as love. Thus, while a Spanish speaker can distinguish between "loving" someone and "caring for" or "being fond of" someone, an English speaker lacks the vocabulary to express that nuance as clearly.

Just as English has one word and Spanish has two, the ancient Greek language—in which the Gospels were written—had three primary words to express different forms of love: *eros, philia*, and *agápē*.

- *Eros* refers to romantic or passionate love, the love that arises between a man and a woman, often not born of reason or choice but something that imposes itself on the human heart.
- *Philia* is the love of friendship, familial affection, or emotional closeness—such as the love between siblings, parents, and children, or between close companions and even pets. It is this kind of love that Paul describes in his First Letter to the Corinthians:
  "Love is patient, love is kind, it is not jealous or boastful, it is not arrogant or rude. Love does not insist on its own way. It is not quick-tempered, nor does it brood over injury. It does not rejoice in wrongdoing but rejoices in the truth. It bears all things, believes all things, hopes all things, endures all things". (1 Corinthians 13:4–7)
- *Agápē*, however, is reserved for divine love—the kind of love Jesus has for us. It is unconditional, self-giving, and expecting nothing in return. It is the highest form of love, and from a human perspective, it often seems almost unnatural. But this is the challenge: to strive toward *agápē*—to love without conditions.

When we translate the word "love" in Peter's dialogue with the risen Jesus (John 21:15–17) into Greek, a much deeper layer of meaning is revealed:

> After breakfast, Jesus asked Simon Peter: — "Simon, son of John, do you *agápē* me more than these?" Peter answered: — "Yes, Lord, you know that I *phileō* (love) you." Jesus said: — "Feed my lambs." Jesus asked him again: — "Simon, son of John, do you *agápē* me?" Peter replied: — "Yes, Lord, you know that I

phileō (love) you." Jesus said: — "Tend my sheep." The third time, Jesus asked him: — "Simon, son of John, do you *phileō* me?" Peter was saddened that Jesus had asked him a third time if he *phileō* (love) Him, and said: — "Lord, you know everything; you know that I *phileō* you." Jesus said to him: — "Feed my sheep." (John 21:15–17, adapted to Greek word distinctions)

In this dialogue, the Lord is inviting Peter to embrace a love far greater than human affection—*agápē*. But Peter responds each time with *phileō*—the kind of love his humanity is capable of offering. On the third attempt, Jesus meets Peter where he is, lowering the demand from *agápē* to *phileō*, acknowledging the limit of Peter's love, yet still entrusting him with the care of His flock.

This exchange illustrates the gentle patience of Jesus, who desires our growth in love, but who also accepts our limitations as the starting point for transformation.

Since returning to my Church, I began searching for the most rational way possible to establish a relationship with God. That is one of the reasons why apologetics so deeply captured my interest. In my effort to rationalize everything I was discovering and learning, I came to realize that the best way to build this relationship was by comparing it to the one I had with my family when I was a five-year-old child.

Even though I am now an adult, before God I am still that same child. He is my Father and my Mother, I am His little son, and all the people around me are my siblings. The relationship I had with my family as a child is the same kind of relationship I seek—and now understand—I have with God.

As a child, I did not understand many things about my parents. I did not always grasp why they did what they did, or why they said what they said. I did not understand why they sometimes gave me food I did not like, or why we had to do things that made me uncomfortable—like going to the doctor or visiting relatives I did not enjoy seeing.

It is the same with God.

As a child, I did not worry about where the food came from, or who paid for the clothes, or how the house was maintained, or where my toys came from. The truth is everything was guaranteed—nothing was ever lacking. All I had to do was say I was hungry, and as if by magic, food appeared before me. It was as though the pantry was endless and self-replenishing.

The clothes they gave me, even if they did not have horses or crocodiles on them, always kept me warm and made me look good.

If I woke up crying in the middle of the night from a bad dream, one of my parents would come to comfort me right away, staying with me until I calmed down and fell back asleep—no matter how tired they were or how early they had to get up the next morning. They were never too exhausted, and they were always awake when I was.

It is the same with God.

If one of my siblings got sick, and I offered them a glass of water or showed concern, it made my parents happy. They knew I could not heal anyone, and that what I did might not change anything—but they were glad to see my love expressed in those small gestures. They celebrated that my heart cared in moments like that.

If I fought with a sibling, it disturbed my parents until we forgave each other. They did not enjoy seeing us quarrel—but they loved it when we hugged, shared, and had fun together.

It is the same with God.

When Mother's Day approached, we were asked at school to bring materials to make a craft and give it to our mothers as a gift. I had to ask my mother to buy the supplies—she was always in charge of those matters—and on more than one occasion, she even helped me finish the project. Yet when I gave it to her, she received it with great surprise, as if she had never seen it before. My parents genuinely enjoyed being asked for things, especially for their help and advice.

They would cover my ears so I would not hear what I should not hear, shield my eyes so I would not see what I was not ready to see, and protect my heart so that emotions did not arise before their time. Even when I believed that all words, images, and feelings were equally good, they knew better. I remember once, my mom became really upset and scolded me when she caught me trying to stick a popsicle stick into an electrical outlet. She warned me that if she caught me doing it again, she would punish me severely. It took me many years to truly understand why she had reacted that way.

It is the same with God.

One of the worst fears I had as a child—the greatest anguish, the deepest dread—was going to the dentist. I was convinced that nothing in the world could be more terrifying. My mother reassured me with a gentle laugh, saying, "Don't worry, everything will be fine." I am not going to let go of your hand for a moment, and it will be over before you know it."

She knew I would be happy again soon, but in my world, it felt like the end of everything. In the end, I did not trust the dentist—but I trusted her. So, I squeezed her hand tightly and surrendered to the ordeal. I did not understand the need for such torture, but something in me believed that if she was by my side, then everything would be okay.

And she was right. It passed quickly; I did not die. In fact, I was better afterward. I could eat without pain again.

It is the same with God.

When I was five, I saw my parents as real-life superheroes—like Superman and Wonder Woman. They never got tired. They could do everything. They could see the future. They did not need sleep or food. I could not lie to them because they always knew the truth. They solved every problem I had with astonishing ease. They were not afraid of anything, and they chased away the monsters that sometimes crept into my room at night.

They were always right. And somehow, they made each of us feel like the favorite child. They were like walking encyclopedias, ready to answer any question. I believed we were richer than Bill Gates because we never lacked food, clothes, toys, or the occasional movie outing or family walk. From my perspective, we had everything in abundance.

It is the same with God.

I have often heard people interpret Matthew 18:3 as referring to a child's innocence or purity of heart: "Amen, I say to you, unless you change and become like little children, you will never enter the kingdom of heaven." (Matthew 18:3)

They say we need to rid ourselves of impure thoughts to become more like children. While that is a valid interpretation, I have taken these words as an inspiration for the kind of relationship with God I have described here.

A child has complete trust in his parents and surrenders to them. He knows that everything comes from them, that they solve all problems, and that without them, he is in trouble. That is why a child clings to their hand—and his world changes.

Now, Matthew 5:3 makes even more sense: "Blessed are the poor in spirit, for theirs is the kingdom of heaven." (Matthew 5:3)

This passage says the same as the previous one: the poor, like children, have nothing and are completely dependent on someone else.

Another passage that fills me with confidence is: "If you then, who are evil, know how to give good gifts to your children, how much more will your Father in heaven give good things to those who ask him?" (Matthew 7:11)

This confirms for me that the kind of relationship I have developed with God—one based on trust, love, and childlike dependence—is not only valid but intimately aligned with the Gospel.

With all the trials and triumphs that King David experienced, I believe he also came to develop this kind of relationship with God. In Psalm 139, he writes:

> Lord, you have probed me and you know me. You know when I sit and when I stand; you understand my thoughts from afar. You sift through my travels and my rest; with all my ways you are familiar. Even before a word is on my tongue, Lord, you know it all. Behind and before you encircle me and rest your hand upon me. Such knowledge is too wonderful for me, far too lofty for me to reach. (Psalm 139:1–6)

Some people remain skeptical of the Bible—not because of what it says, but because of what others have said about it. Many view religion as a tool of control, claiming that the Bible was manipulated by the Church to instill fear of eternal punishment and thus maintain power.

Some argue that biblical authors fabricated stories to label certain behaviors as "sinful" to create guilt and obedience. Others point out that the Gospels were written years after the events and question their reliability. There are even those who believe that "priests" of old deliberately excluded certain writings to shape the Bible into a tool of fear and control.

These claims are numerous—and come from many places and perspectives.

In my book *What You Wanted to Know About the Catholic Church but Were Afraid to Ask*, I dedicated several chapters to refuting these and other claims. For that reason, I will not go into detail here, repeating the facts and evidence that disprove these slanders. Most of them arise from a lack of historical knowledge, and even worse, from a lack of logic and common sense.

With the information presented in this book, I believe it has become sufficiently clear that the prophets spoke on behalf of God. Only under divine inspiration could they have prophesied what they did, with those prophecies being fulfilled to the letter, centuries later.

Would it be reasonable to believe that all these men, who enjoyed a close friendship with God, betrayed Him in the end by lying in their writings? All of them?

And if those writings were nothing more than human inventions, why would Jesus Himself have repeated them to instruct, correct, and guide the people—calling them the *Word of God*—if they were not?

As for the New Testament, what would lead us to believe that the apostles, who had seen the risen Jesus (thus confirming that He was truly God), would put words in His mouth that He never spoke?

I was barely eight years old when my father navigated a long curve near the top of a mountain. Suddenly, on the horizon, the Atlantic Ocean appeared, filling the entire landscape in front of us. Even now, as I write these lines, I can vividly recall that moment: the car we were in, the joy we felt when my mother pointed to that vast bluish expanse and told us it was the sea.

I do not remember what I was wearing, or whether I was in the front or back seat. I do not remember the exact words they used to introduce us to the ocean. But what I do remember, and what has stayed with me for a lifetime, was the joy of that first sighting—the excitement, the noise, the celebration my siblings and I shared in that unforgettable family moment.

Now, if I were to write about that day, I would describe it much as I have here.

Would it change the story if I wrote that my mother said, "Look, kids, that's the sea" instead of "There you have the famous sea" or "Behold, the Atlantic Ocean"?

Would the truth of the story be invalidated if it turned out that it was my father who said it, and not my mother, as my memory recalls?

Of course not.

The essence of the story is what matters: that we were in the car, as a family, that we saw the ocean for the first time, that someone pointed it out to us, and that it brought us immense joy—a memory we still treasure.

That is exactly what the apostles did.

They wrote down the essence of Jesus' life: His teachings, His miracles, His thoughts, His warnings, His advice, His commands, and His promises.

With the theses presented in this work, I believe you can feel confident that none of the disciples dared to distort his words or lie when they wrote the testimony of all their experiences with the Master; testimonies that, in the long run, cost them their lives.

Moreover, the apostles were not the only witnesses. Jesus was almost always surrounded by crowds, and these people served as living guarantors of what was later written by the New Testament authors.

As I have emphasized throughout this book, the Bible possesses certain characteristics that are unmatched by any other sacred text belonging to fully established religions such as Islam, Hinduism, or Buddhism.

The Bible is the only sacred text that contains a narrative of the creation of the universe that aligns remarkably well with the most recent scientific discoveries. Moreover, it is the only one that includes prophecies foretelling that God would take on human form and dwell among us.

I have demonstrated how, from a mathematical standpoint, it is impossible for those prophecies to have been fulfilled in the life of anyone other than the Messiah. So—do you now know how to respond when someone questions your certainty of belonging to the true faith?

The God of the Bible is our Creator, and Jesus confirmed it through His works, His life, and above all, His resurrection.

I presented the great dilemma that arises when we take all of Jesus' words seriously. Of course, He said many things that were beautiful, altruistic, and full of hope—words that appeal to both believers and non-believers. That is why some are content to see Him merely as a good man, while conveniently ignoring—out of ignorance or preference—that He also claimed to be God.

But this is precisely the issue: if that claim is true, then Jesus is the Messiah, God made man. If it is false, then He is not simply a mistaken teacher—He is a madman.

Do you see the enormity of this question?

If you are not yet firmly convinced that Jesus rose from the dead, but you admire His teachings, remember this: those teachings would then be the words of a delusional man. That is why I have placed such importance in this work on providing a broad and compelling body of evidence supporting the historical reality of the resurrection.

There is no other burial in antiquity about which we have as much information as we do about the burial of Jesus. This has enabled me to compile the thirteen theses presented in this book—though there are many more.

Of all the topics explored in Christian literature, none has received more attention than the resurrection of Jesus. Entire treatises have been devoted to it, including:

- *The Resurrection of Jesus*, by James Orr
- *The Resurrection of Our Lord*, by William Milligan
- *The Resurrection and Modern Thought*, by W. J. Sparrow-Simpson

The resurrection of Jesus was not the result of the apostles' faith. On the contrary, it was the resurrection that gave rise to their faith. And it is this very faith that has been passed down to us today through Sacred Scripture, Sacred Tradition, and through the apostolic succession of bishops.

Those of us who belong to the Catholic Church have become so accustomed to the figure of Jesus and His teachings that we often remember the great event of His resurrection only during Holy Week. We regard it as just one more episode among the many from His three years of public ministry.

But the Master—who had to explain the meaning of the resurrection to His own disciples (cf. Luke 24:13–35)—and Paul, who declared that "if Christ has not been raised, then our faith is in vain" (cf. 1 Corinthians 15:14), both emphasized the immense significance of that empty tomb.

Without the resurrection of the Lord:

- We would still be sacrificing lambs to seek the forgiveness of sins.
- We would still be circumcising as required by the Law.
- We would still be bound to the over six hundred Mosaic commandments.
- We would still be forced to let the sick suffer on the Sabbath, since healing would violate the Law.

But Jesus rose—and with that, He breathed new life into the Law.

The sacraments became visible signs of God's invisible grace, made present through the Holy Spirit, as Jesus had promised before ascending into heaven.

He remained present in body through the Eucharist, to nourish us.

And He remained present in spirit, to guide us to the Father on our pilgrimage toward eternal life.

If you are one of those people who—for whatever reason—do not read the Bible regularly, I warmly invite you to do so. After reflecting on the topics covered in this book, I trust you will agree with me that the Bible can be seen as an autobiography of our Father. And truly—who would not want to read the autobiography of one of their earthly parents, if such a thing existed?

Even simple curiosity would be motivation enough. But beyond that, the desire to know more about the one who brought us into the world and has loved us so deeply should be the greatest reason of all to open it—and to keep coming back to it.

I have already given you the assurance that the words found in today's Bible are the same ones written by the prophets hundreds of years ago. You can now confidently set aside the myths and excuses that may have kept you from reading it.

If you are unfamiliar with the Scriptures, I recommend that you begin with the Gospel of Luke, then continue with the Gospel of Matthew or Mark, followed by the Letters of James and John, then the Letter to the Romans and the Psalms. In this way, little by little, you will go deeper and deeper into the Word of God.

Do not worry about understanding every word or finding a message in each sentence. Simply approach the Scriptures to spend time with your best friend—because that is exactly what God wants to be for you.

I come now to the end of this work with the hope that I have helped you confirm your Catholic faith on rational grounds, supported by science, mathematics, history, and logic.

The Doctor of the Church, Saint Anselm of Canterbury (11th century), taught that it is necessary to believe to understand, and then to strive to understand what we believe. According to him, to begin without faith is presumptuous, but to avoid invoking reason afterward is negligent.

Today, more than ever, we must take that second step recommended by this illustrious doctor of the Church. Atheist and agnostic ideologies continue to attract new followers each day, often by using false scientific narratives and misleading questions—questions that someone better prepared could refute with the kind of facts and reasoning presented in this book.

APPENDIX A

WHO IS GOD AND WHO IS NOT GOD?

One of the first prayers I learned in childhood was the Apostles' Creed. A particular phrase always stood out to me: "[...] On the third day He rose from the dead, ascended into heaven, and is seated at the right hand of God the Father almighty [...]"

Whenever I recited those words, I would picture a vivid scene in my mind: two majestic figures—Father and Son—each dressed in white tunics with golden sashes, standing on a massive cloud, surrounded by other smaller clouds and plump angels playing instruments that looked like miniature harps.

They were seated on golden thrones, adorned with precious stones, gazing down toward Earth. The scene was set in a space above—a brilliant blue expanse, high above the world.

I imagined the Father like a kind of Santa Claus figure: a bit overweight, with blue eyes, gray hair, a gentle smile, and white skin. In my imagination, His throne was larger and more imposing than that of the Son.

And the Son, seated at His right hand, looked—at least in my childhood imagination—like the English actor Robert Powell, who portrayed Jesus in Franco Zeffirelli's[160] 1977 film *Jesus of Nazareth*. For some reason, even though Powell had brown hair in the film, I imagined Jesus with gray hair.

Already in my adulthood, through conversations with others, I have come to realize that many people share the same childhood image of God and His Son after the ascension to heaven. Just as the radio leaves it to our imagination to give a face and a body to the voice that emerges from the speaker, we tend to do the same with God. We want to see Him, to picture His face, to assign Him a form.

The Gospel of Matthew tells us: "And a voice from heaven said, 'This is my beloved Son, with whom I am well pleased.'" (Matthew 3:17)

But it leaves the question open: What does God look like? What is His face and body like? Who is He?

In his novel *The Shack*, William Paul Young explores a deeply personal and spiritual encounter between a man grieving the death of his young daughter—murdered by a serial killer—and the Holy Trinity. One of the book's most provocative elements is how the author represents each Person of the Trinity: The Father appears as an African American woman, The Holy Spirit as an Asian woman, and the Son as a Jewish man.

The story was adapted into a film in 2017, with Octavia Spencer[161] portraying the Father—affectionately called "Papa." The creative portrayal stirred conversation but also illustrated how deeply rooted our desire to visualize God truly is. The question of what God looks like has captivated the human imagination for as long as we have historical memory.

It is the same kind of curiosity we have when getting to know any person of significance.

---

[160]Gian Franco Corsi Zeffirelli was an Italian film director, producer, and designer renowned for his extensive body of acclaimed films, operas, and theatrical productions.

[161]Octavia Spencer (born May 25, 1970, in Montgomery, Alabama) is an American film and television actress, director, producer, and screenwriter. She has received numerous accolades, including an Academy Award, a Golden Globe, a BAFTA, and three Screen Actors Guild Awards.

For instance, if someone were to ask, "Who was Gabriel García Márquez?" we might answer: He was a Colombian writer, winner of the Nobel Prize in Literature (1982), author of iconic novels such as *One Hundred Years of Solitude, Chronicle of a Death Foretold,* and *In Evil Hour.* He also wrote short stories like *The Incredible and Sad Tale of Innocent Eréndira and Her Heartless Grandmother* and *Big Mama's Funeral,* among others. He lived to be 87 years old, and passed away in Mexico City, where he had spent his final years, after battling lymphatic cancer.

If the question is, "What was he like?"

We could describe him as a brilliant and humorous man, someone who brought laughter to his friends, a tireless seeker of peace, and although openly socialist, someone who kept his distance from political entanglements. He remained faithful to his ideals and regarded friendship as one of life's most precious gifts—something to be cherished and defended at all costs.

As for his physical appearance, we might say he had curly hair, a round face, high cheekbones, bushy eyebrows, and a full mustache that highlighted his perpetual smile. He had a broad, high forehead, a hooked nose, and light skin that contrasted with his dark eyes.

These are the kinds of questions we naturally ask when we want to truly know someone: What is their name? Where are they from? What do they look like? What are they like? What did they accomplish? What legacy did they leave behind?

So, it is only natural that we would want to ask the same things about God.

WHAT IS THE NAME OF GOD?

After Moses left behind his adoptive royal family and fled the land of Egypt, he sought refuge in the region of Midian. There, he met Zipporah, the daughter of Jethro, whom he later married. In that land, Moses took on the humble task of tending sheep, caring for the flocks of his father-in-law.

Forty years later, while tending the flock on Mount Horeb, he saw a strange sight: a bush that burned but was not consumed. As he approached, a voice called his name and told him to remove his sandals, for the ground he was standing on was holy. The voice identified itself as the God of his ancestors—the God of Abraham, Isaac, and Jacob—and entrusted him with a mission: to lead the people of Israel out of Egypt and bring them to the land of freedom.

Before accepting the mission, Moses—like one who needs to be sure of all the details—posed a final, crucial question:

> But, said Moses to God, "when I go to the Israelites and say to them, 'The God of your ancestors has sent me to you,' they may ask me, 'What is his name?' What then shall I say to them? (Exodus 3:13)

Moses, who had lived in a culture that worshiped many gods—the gods of the sun, fire, moon, and death—wanted to know which God was sending him. Among so many deities, who was this one? The reply came:

> I AM WHO I AM. This is what you are to say to the Israelites: 'I AM has sent me to you.' (Exodus 3:14)

Some translations render it: "I AM THAT I AM." Others phrase it as "The ONE WHO IS."

What God gave Moses was not a name in the way we understand names—like "chair," "moon," "Carlos," or even the names of Egyptian deities such as *Ra* (god of the sun), *Amun* (king of the gods), *Thoth* (god of the moon), or *Hathor* (goddess of love and joy). "I AM" was not a name—it was a revelation of His nature: He IS.

This declaration points not to what God does, but to who God is—existence itself.

So, how are we to refer to Him if He did not give a proper name?

In ancient Hebrew, vowels were not written, only consonants. The name revealed in the Pentateuch appears as the four consonants: Yod–Heh–Vav–Heh (יהוה), which was pronounced as *Yahweh* (*Iahuéj*).

When translated into Latin, these letters became YHWH, and later, in English, the form *Yahweh* was used.

In the Middle Ages, the Masorete Jews (who replaced the scribes of Jesus' time) took the vowels from the words *Elohin*, meaning "Mighty God," and *Edonay*, meaning "The Lord," and mixed them with *Yahweh*.

From this combination came the hybrid word *YeHoWiH*, which later evolved into *Jehovah*, a form adopted in many Protestant Bible translations to refer to God.

But it is important to remember: These are names created by human beings. They are not the revealed name of God.

WHO CREATED GOD?

This is a question that has been, is being, or will be asked by many people who believe in God or a Creator. Who created the Creator? Who created God? The question may seem valid. Syntactically, it is well-formed. But it does not make sense.

Not all grammatically correct questions are logically valid. For example: "Do you remember what you ate yesterday?" — This is both grammatically correct and logical. But "Do you remember what you died yesterday?" — While structurally identical, this question is nonsensical. The phrase "what you died" lacks semantic coherence.

Here are more examples of grammatical absurdities:

- "How many meters are in a liter of water?" — Syntactically correct, but illogical since volume cannot be measured in meters.
- "How do I not forget places I've never been?"
- "What does a triangle with four angles look like?"[162]

---

[162]In his *Summa contra Gentiles*, Saint Thomas Aquinas wrote: "Since the principles of certain sciences—such as logic, geometry, and arithmetic—are derived solely from the formal principles of things, upon which their very essence depends, it follows that God cannot perform actions contrary to these principles. For example, from a species, the genus cannot be predicated; it is impossible that lines [radii] drawn from the center of a circle are not equal, just as it is impossible that the sum of the internal angles of a triangle does not equal two right angles."

These are all examples of contradictory or impossible ideas, masked by grammatically valid phrasing.

So, when someone asks[163], "Who created **G**od?", what they are really asking—unknowingly—is, "Who created the uncreated Creator?"

And the only logical answer is: If anything created that "**g**od," then that creator would be the real **G**od.

In other words, the God of Abraham, Isaac, and Jacob—the **G**od Jesus called Father, the Creator described in Genesis—was not created. That is precisely what defines Him as **G**od: He simply IS. "I AM WHO I AM" (Exodus 3:14)

He did not give Moses a name in the ordinary sense—like chair, table, Carlos, Ra, or Toth.

"I AM" is not a name, but a revelation of His nature: He exists by necessity, not by creation.

Therefore, asking "Who created **G**od?" is a contradiction—it assumes that **G**od is a contingent being, like everything else in creation, when by definition, He is necessary.

All things that have existed, exist now, or will exist in the future share a property called contingency. This means they might exist or might not.

I am contingent—I exist, but I might not have existed. The door of your house is contingent—it exists, but you could have a different one. The sun is contingent—if it did not exist, we would not either, but the rest of the universe might.

God, on the other hand, is not contingent. He does not depend on anything for His existence. He is eternal and self-existent. He is necessary—the opposite of contingent.

---

[163]In Christianity, the term **god** (in lowercase) is often used as a synonym for "idol." In this paragraph, I have emphasized the distinction by using bold type to highlight the initial letter—God (uppercase) refers to the one true God in Christian belief, while **god** (lowercase) denotes a false deity or idol.

In his exposition of the Five Ways to prove the existence of God, Saint Thomas Aquinas refers to God as the First Cause[164]. For there to be contingent beings like us, there must be a non-contingent being that caused them all. If we exist, then the One who must exist—God—cannot not exist.

So, when someone asks, "Who created God?," they are unknowingly trying to apply the rules of contingency to a being that is necessary. It is like asking: "What is the tastiest food you've never tasted?" or "How do I remember something that never happened?"

These are self-contradictory questions. The same applies to "Who created God?"

It is important to be cautious with questions that contain built-in contradictions because they can unsettle believers who are not prepared to see the logical error in them.

One classic example is: "Can God create a stone so heavy that He Himself cannot lift it?"

This question is not a genuine test of omnipotence. It misunderstands what omnipotence means. God's omnipotence does not include the power to do the logically absurd—like making a square with three angles or a living dead person. These are not real things; they are conceptual contradictions.

The same goes for the immovable stone. The question is designed not to explore truth, but to confuse and cast doubt. Once we understand the logic behind omnipotence, the malicious intent of the question is easily disarmed.

The same can be said of the question: "What was God doing before the creation of the universe?"

Saint Augustine famously quipped: "He was preparing hell for those who ask such questions."

---

[164]Long before Saint Thomas Aquinas, Aristotle—writing in the 4th century BC—was among the first philosophers to speak of a primary cause.

But in theological terms, the question cannot be answered, because time itself was created when God created the universe. The concept of "before" only makes sense within time—and time did not exist until God willed it into being.

WHAT IS GOD PHYSICALLY LIKE?

Yuri Gagarin was the first human being to travel into space. He did so on April 12, 1961, aboard the Russian spacecraft Vostok 1. Sometime later, the then General Secretary of the Communist Party, Nikita Khrushchev[165], addressed the plenary session of the committee with the following declaration: "Gagarin flew into space, but he saw no God there."

I too, at a certain point in my life, shared that illusion—the hope that one day, through a super telescope or the report of a lucky astronaut, we might be able to see God, to learn something about His appearance or anatomy.

But Khrushchev's comment—though delivered with the confidence of a world leader—was intended to be a conclusive argument against the existence of God.

Greek mythology[166] remains one of the most familiar systems of ancient belief. The Greeks had gods with names and forms: Zeus, Cronus, Poseidon, Uranus, Hades, Eros, and many others. After their victory over the Titans, the world was divided: Zeus took the sky and air, Poseidon the waters, and Hades the underworld.

These gods not only had human names and shapes—they acted like humans. They married, gave birth to other gods, and often intervened in human affairs. They were prone to jealousy, rage, desire, betrayal, and

---

[165]Nikita Khrushchev led the Soviet Union during part of the Cold War, serving as First Secretary of the Communist Party of the Soviet Union from 1953 to 1964.

[166]Greek mythology is the collection of myths and legends from the culture of Ancient Greece, centered on its gods, heroes, the nature of the world, and the origins and significance of its religious practices and rituals. The term Ancient Greece refers to the historical period extending from the Greek Dark Ages (beginning around BC 1200, following the Mycenaean collapse and the Dorian invasion) to BC 146, marked by the Roman conquest after the Battle of Corinth.

pride—sharing the same weaknesses and passions as the mortals who worshiped them.

Given that legacy, it is not surprising that many Christians wonder about the physical form of God. As I mentioned earlier, I did so frequently before my understanding of the Father matured.

What we know of God is only what He has revealed to us. Regarding His nature, Jesus tells us: "God is Spirit, and those who worship him must worship in spirit and truth." (John 4:24)

God is not a natural being, but supernatural. That is why Catholic tradition avoids referring to Him simply as a "being", or as some supreme entity above all others. Rather, He IS.

In the Latin of St. Thomas Aquinas, God is defined as *ipsum esse subsistens*—the very act of being itself. God has existence by His own essence, unlike all other creatures who receive their being from another.

According to Hebrew grammar, the phrase "I AM WHO I AM" (Exodus 3:14) can also be translated as "I am the one who was, who is, and who will be." In other words, He who exists by Himself—uncreated, eternal, necessary.

In Genesis, we read: "Then God said, 'Let us make man in our image, after our likeness.'" (Genesis 1:26)

Does this mean that God has arms, legs, eyes, and other human features?

The answer is: No. We were made in His image because He infused us with a soul in His image. This is revealed in: "Then the Lord God formed man from the dust of the earth and blew into his nostrils the breath of life, and the man became a living being." (Genesis 2:7)

We can create because He is Creator. We can love because He is love. We can forgive because He is forgiveness. We can be faithful because He is faithfulness. We can be patient because He is patience.

These are not physical traits. They are spiritual and moral reflections—manifestations of our soul, made in the image of God, and they are what set us apart from the rest of creation.

This image of God is expressed in three key dimensions:

- Mental likeness: We were created with intellect and free will, capable of reasoning and choosing. Every time someone writes a poem, composes music, solves a problem, or acts creatively—they reflect the mind of God.
- Moral likeness: We were created with an initial state of justice and innocence, a reflection of God's holiness. Whenever someone tells the truth, pursues justice, or turns from wrongdoing, they manifest God's moral image.
- Social likeness: We were created for relationship, which reflects the Trinitarian nature of God. The first relationship was between God and man, and then God created woman, because: "It is not good for the man to be alone." (Genesis 2:18). Every time someone loves, cares, embraces, marries, helps a neighbor, or prays, they are proclaiming the image of God within themselves.

Let us not forget that God is One and Triune: Father, Son, and Holy Spirit—three Persons in one true God.

In Genesis 18:1–2, God appears to Abraham by the oak of Mamre, and three figures stand before him: "The Lord appeared to him by the oak of Mamre, as he was sitting at the entrance to his tent during the hottest part of the day. Looking up, he saw three men standing nearby."

On this occasion, Abraham and Sarah were the only ones to witness this mysterious encounter. But later, thousands of people would see the Son, when He became incarnate as the child of Joseph and Mary and lived among us for about thirty-three years: "And the Word became flesh and lived among us, and we have seen his glory, the glory as of the Father's only Son, full of grace and truth." (John 1:14)

The Holy Spirit also took on visible form on various occasions: As a dove at Jesus' baptism (cf. Matthew 3:16) and as tongues of fire at Pentecost (cf. Acts 2:3). But we must remember: these forms are not their essence.

They are chosen manifestations, temporary and symbolic, used by the Triune God to relate to humanity. They serve a specific purpose, not to define His nature, but to express His love in ways we can understand.

WHAT ARE THE CHARACTER AND INTELLECT OF GOD LIKE?

The evangelist John, in his first letter, declares: "Whoever does not love does not know God, for God is love." (1 John 4:8)

Notice that John does not say that God has a lot of love, or that He is the greatest expression of love, or even that He possesses the highest form of love ever known. No—he says simply and profoundly: God IS love.

Take a moment to absorb what this means: He is love itself.

We also know from Scripture and tradition that God is infinite—without limit. Everything in creation, no matter how vast or powerful, is finite. The seas contain a measurable amount of water. The energy in an atom has a quantifiable magnitude. Even the heat of the sun has an upper limit.

But God has no limits—in any sense.

To say that God is infinitely perfect means that there is nothing good, desirable, or valuable that He does not possess in an absolutely unlimited degree. In theology, we say that the perfections of God are identical with His essence—they are what He is. This means that, to be precise, we should not say "God is good," but "God is Goodness"; nor "God is wise," but "God is Wisdom," etc.

God is also described in Scripture with many other divine attributes: Omniscient – He knows everything (cf. Psalm 139:1–16), Benevolent – He desires only what is good (cf. 1 John 4:8), Omnipotent – He can do all things (cf. Job 40:1), Omnipresent – He is everywhere at once (cf. Psalm 139:7–10), Immutable – He does not change (cf. Psalm 102:27; Revelation 1:8), One and Unique – There is no other like Him (cf. Deuteronomy 32:39; Isaiah 45:5).

All Scripture reveals to us a loving Father: One who knows us deeply, Who is always with us, Who loves us unconditionally, no matter what.

In January 1999, Barry Adams[167] published a project online[168] called Father's Love Letter[169]. He compiled passages from the Holy Scriptures into a personal message from God the Father to His children. Here is that letter:

> My Child, you may not know me, but I know everything about you (Psalm 139:1). I know when you sit down and when you rise up (Psalm 139:2). I am familiar with all your ways (Psalm 139:3). Even the very hairs on your head are numbered (Matthew 10:29–31). For you were made in my image (Genesis 1:27). In me you live and move and have your being (Acts 17:28), for you are my offspring (Acts 17:28). I knew you even before you were conceived (Jeremiah 1:4–5). I chose you when I planned creation (Ephesians 1:11–12). You were not a mistake, for all your days are written in my book (Psalm 139:15–16). I determined the exact time of your birth and where you would live (Acts 17:26). You are fearfully and wonderfully made (Psalm 139:14). I knit you together in your mother's womb (Psalm 139:13) and brought you forth on the day you were born (Psalm 71:6). I have been misrepresented by those who don't know me (John 8:41–44). I am not distant and angry, but am the complete expression of love (1 John 4:16), and it is my desire to lavish my love on you (1 John 3:1), simply because you are my child and I am your Father (1 John 3:1). I offer you more than your earthly father ever could (Matthew 7:11), for I am the perfect Father (Matthew 5:48). Every good gift that you receive comes from my hand (James 1:17), for I am your provider and I meet all your needs (Matthew 6:31–33). My plan for your future has always been filled with hope (Jeremiah 29:11), because I love you with an everlasting love (Jeremiah 31:3). My thoughts toward you are countless as the sand on the seashore (Psalm 139:17–18), and I rejoice over you with singing (Zephaniah 3:17). I will never stop doing good to you (Jeremiah 32:40), for you are my treasured possession (Exodus 19:5). I desire to establish you with all my heart and all my soul (Jeremiah 32:41), and I want to show you great and marvelous things (Jeremiah 33:3). If you seek me with all your heart, you will find me (Deuteronomy 4:29). Delight in me and I will give you the desires of your heart (Psalm 37:4), for

---

[167] Co-founder of Father Heart Communications and associate pastor of Westview Christian Fellowship.

[168] www.FathersLoveLetter.com

[169] *Father's Love Letter*, by permission of Father Heart Communications © 1999 www.FathersLoveLetter.com

it is I who gave you those desires (Philippians 2:13). I am able to do more for you than you could possibly imagine (Ephesians 3:20), for I am your greatest encourager (2 Thessalonians 2:16–17). I am also the Father who comforts you in all your troubles (2 Corinthians 1:3–4). When you are brokenhearted, I am close to you (Psalm 34:18). As a shepherd carries a lamb, I have carried you close to my heart (Isaiah 40:11). One day I will wipe away every tear from your eyes and take away all the pain you have suffered on this earth (Revelation 21:3–4). I am your Father, and I love you even as I love my Son, Jesus (John 17:23), for in Jesus, my love for you is revealed (John 17:26). He is the exact representation of my being (Hebrews 1:3). He came to demonstrate that I am for you, not against you (Romans 8:31), and to tell you that I am not counting your sins (2 Corinthians 5:18–19). Jesus died so that you and I could be reconciled (2 Corinthians 5:18–19). His death was the ultimate expression of my love for you (1 John 4:10). I gave up everything I loved that I might gain your love (Romans 8:31–32). If you receive the gift of my Son Jesus, you receive me (1 John 2:23), and nothing will ever separate you from my love again (Romans 8:38–39). Come home and I'll throw the biggest party heaven has ever seen (Luke 15:7). I have always been Father and will always be Father (Ephesians 3:14–15). My question is... Will you be my child? (John 1:12–13). I am waiting for you (Luke 15:11–32). Love, Your Dad, Almighty God

With all this in mind, it becomes much easier to appreciate that beautiful phrase spoken by Jesus and recorded by the evangelist Matthew: "If you then, who are evil, know how to give good gifts to your children, how much more will your heavenly Father give good things to those who ask him?" (Matthew 7:11)

WHAT ARE THE WORKS AND LEGACY OF GOD?

Understanding the Creator as the One capable of making something out of nothing—which is, in fact, the very definition of creating (as opposed to making, producing, transforming, or converting)—then the works of God include the entire visible and invisible universe. That is, all of Creation, including you.

The heavens proclaim the glory of God, and the firmament displays the work of his hands. Day unto day pours forth speech; night unto night imparts knowledge. There is no speech, no words, no voice that can be heard. Yet their message reaches the entire world, their words go out to the ends of the earth. He has set a tent for the sun in the heavens, which comes forth like a bridegroom from his bridal chamber, and like a champion sets out to run his course. It rises from one end of the sky and completes its circuit to the other; nothing is hidden from its heat. (Psalm 19:1–6)

Have you ever been invited to someone's home—someone you have never met—and, upon entering, you begin to form an impression of the host simply by observing the surroundings?

Whether the house is clean or messy, what kind of art hangs on the walls, what music plays on the radio, which books fill the shelves, what is on the television—all these things reveal something about the person who lives there.

The same can be said of God. We are, after all, honored guests in His house—this magnificent home called Earth.

So, what can we say about Him by looking around us?

We can say, without hesitation, that He is incredibly generous. He made everything in abundance: The seas, the stars, the snow, the trees, the birds, the colors, the scents, the flavors—all these are provided lavishly. Even something as rare and expensive as diamonds are still being found today, despite centuries of mining. Abundance is written into the fabric of Creation.

We can also say that He is limitlessly creative. Just consider the sheer variety in nature: Ants, stars, elephants, octopuses, comets, whales, snowflakes, eagles, waterfalls, caves, dragonflies, oceans, worms, tigers, fruits, roses, emeralds, volcanoes, parrots, plants, stalactites, trees, glaciers, butterflies, rivers, rain, dogs, and—most astonishing of all—man.

This diversity of forms, abilities, sizes, movements, means of nourishment, reproduction, adaptation, contribution, destruction, illumination, absorption, and expulsion—His creativity surpasses any list you could make.

We can also say that God is immensely patient. Just think of a star, which takes millions of years to form, all so that we might one day look up at the sky and say, "What a starry night!"

He clearly loves variety. He did not create just one kind of fish to feed us—though that would have been enough—but millions. He did not create just one type of tree to clean the air and give us wood—but countless varieties. Even with apples, there are hundreds of kinds.

And what about humanity?

Even though we all have eyes, a nose, a mouth, ears, hair, skin, and facial structure, we do not find two identical faces. No two voices are the same. No two fingerprints match. Each one of us is uniquely crafted, bearing His image, but reflecting it differently.

This is part of the marvelous legacy of God: A world created in beauty, in abundance, and in love, so that we may know Him through His works, and walk humbly in awe of His generosity, creativity, patience, and goodness.

WHO IS NOT GOD

God is not one of those mythical deities imagined by ancient writers, commonly found in what we now call mythology—gods who shared the same flaws and passions as human beings. These deities were often portrayed as jealous, angry, envious, and vengeful, but also as capable of love, generosity, and compassion. Their moods shifted depending on the moment, and they were frequently depicted as bored wanderers on Earth, sometimes seduced by women, betraying their celestial consorts.

God is not a violent warrior who imposes truth through wars and destruction, as some extremist ideologies portray Him—claiming that violence is divinely sanctioned.

God is not a cosmic police officer, hiding behind us, waiting to catch us in wrongdoing and immediately punish or correct us, like a stern earthly parent watching over our every move.

God is not a force of energy diffused through nature, feeding our bodies and souls, as some pantheistic or spiritualist perspectives suggest.

God is not a puppeteer who manipulates our lives for amusement—sending punishments in the form of illness, failure, or ruin when we misbehave, or dispensing rewards such as health, wealth, fame, or power when we behave.

God is not a narcissist, requiring constant praise and worship to be satisfied, as though His love for us were conditional on our level of devotion.

God is not the so-called "god of the gaps"—a convenient placeholder for phenomena we do not yet understand, such as rain, fire, or eclipses. As scientific understanding fills these gaps, this kind of "god" inevitably diminishes until disappearing altogether.

God is also not a vague blend of all these mistaken concepts—assembled in our minds and hearts based on life experience, cultural background, or upbringing. Unfortunately, when our understanding of God is immature or misinformed, these flawed images can lead us astray, distorting our journey toward the true image of the Father that Jesus revealed—especially in the parable of the prodigal son.

> A man had two sons. The younger said to his father, 'Father, give me the share of the estate that will come to me.' So, the father divided his property between them. A few days later, the younger son gathered all he had and went off to a distant country, where he squandered his wealth in a life of debauchery. After he had spent everything, a severe famine struck that country, and he began to feel the pinch. So, he hired himself out to a local inhabitant who sent him to his farm to tend the pigs. He longed to fill himself with the pods the pigs were eating, but no one offered him anything. Then he came to his senses and said, 'How many of my father's hired workers have more than enough to eat, and here I am starving to death! I will set out and go back to my father and say to him: "Father, I have sinned against heaven and against you. I am no longer worthy to be called your son; treat me as one of your hired workers."' So, he set out to return to his father."
>
> "While he was still far off, his father caught sight of him and was filled with compassion. He ran to his son, embraced him, and kissed him. Then the son said, 'Father, I have sinned against heaven and against you. I am no longer worthy to be called your son.' But the father ordered his servants, 'Quickly, bring out a robe—the best one—and put it on him. Put a ring on his finger

and sandals on his feet. Bring the fattened calf and kill it. Let us feast and celebrate, for this son of mine was dead and has come back to life; he was lost and has been found.' And they began to celebrate."

"Meanwhile, the older son was in the field. As he came near the house, he heard music and dancing. He called one of the servants and asked what was happening. The servant replied, 'Your brother has returned, and your father has killed the fattened calf because he has received him back safe and sound.' The older brother became angry and refused to go in. His father came out and pleaded with him. But he answered his father, 'All these years I have worked for you like a slave and never disobeyed your orders. Yet you never gave me so much as a young goat to celebrate with my friends. But now that this son of yours returns after squandering your wealth with prostitutes, for him you kill the fattened calf!' The father replied, 'My son, you are always with me, and everything I have is yours. But we had to celebrate and rejoice, because this brother of yours was dead and has come back to life; he was lost and has been found.' (Luke 15:11–32)

CONCLUSION

Do you think I have not answered clearly who God is?

The truth is that all our words and concepts fall short of explaining who He truly is. God is, by essence, mystery—a word that comes from the Greek *muein*, meaning to shut one's mouth. Saint Augustine once said: "If you understand it, that is not God."

It is easy for us to understand the world around us—what exists in time and space. We know that if Juan is over there, he cannot be here at the same time. He is Carlos; therefore, he is not Roberto. A table is not a chair. That is a mountain, and that is a bird.

From childhood, we learn to identify and name things through comparison and contrast. We group similar things under one word, and we distinguish them by what sets them apart. This is how we come to know and name everything in our world.

But with God, we cannot do the same.

We cannot say, "There's a table, there's a wall, Carlos is there, I am here—and God is over there." Even St. Thomas Aquinas refused to classify God under any genus—He is not animal, vegetable, mineral, or even some divine category. There is no such genus. Not even the angels share in His nature. They have their own essence, distinct from ours—and God's is different from all of them.

God is not one more "something" among all the things in the world or universe. He is not even the greatest "something." We can say this building is bigger than that one; the Earth is bigger than all the buildings; the galaxy is bigger than the Earth; and the universe is greater than the galaxy. But God is not even the greatest thing in the universe.

God simply IS.

Since God is infinite and perfect, no created being can fully comprehend His nature. He is utterly different from anything that exists or has existed. He is incomprehensible, inaccessible to our finite, limited minds.

As Saint Paul says:

> He is the blessed and only Sovereign, the King of kings and Lord of lords, the only one who possesses immortality and dwells in unapproachable light, whom no one has seen or can see. To him be honor and eternal power. Amen. (1 Timothy 6:15–16)

Let me ask: do you fully understand your spouse or partner?

Even with a lifetime of shared experiences, common language, and communication, we often find that we cannot fully grasp the heart, thoughts, or mysteries of the person beside us—someone of flesh and blood. If we struggle to understand another human being, how much more should we expect to grasp the essence of God?

And yet, the fact that we do not fully understand someone does not mean we cannot love them deeply, faithfully, and completely. We choose to make a life with them not because we comprehend them perfectly, but because we trust, cherish, and love them.

The same should be true with God. But there is one key difference: we do not need to conquer Him. Instead, we need to let ourselves be conquered—by His love.

Like the father in the parable of the prodigal son (Luke 15:11–32), God is always the one who runs to meet us, embraces us, and celebrates our return. He does not wait at a distance. He moves toward us.

APPENDIX B

SOME MATH

I have considered it important to include a brief introduction to numerical notation, to help contextualize what I mean by "large numbers", and to provide a very short overview of the fascinating world of probabilities.

In addressing the first major question of this book, I found myself needing to reference extremely large numbers. However, I realize that not everyone finds it easy to grasp just how big a big number really is— or, conversely, how small a very small number can be.

Clarifying this distinction is the purpose of this appendix.

BIG NUMBERS

First, we must talk about scientific or exponential notation, which is a way of writing large numbers in an abbreviated form. For example, we can write the number one hundred million in the conventional way as: 100,000,000, or in scientific notation as: 1×10^8, which reads "one times ten to the eight."

We can also express 100,000,000 as:

10×10^7, or

100×10^6, or

$1,000 \times 10^5$, and so on[170].

From this, we can deduce that a number written as $m \times 10^{\square}$, where m is called the *mantissa*[171] and e is the *order of magnitude*, is equivalent to writing the number m followed by e zeros to its right.

This notation is also used for very small numbers. For example, one millionth of a unit is equivalent to dividing one unit into a million parts. That would be written as 0.000001, or in scientific notation, 1×10^{-6}. In this case, the exponent indicates the number of zeros to the left of the mantissa.

Here are some examples:

$500 = 5 \times 10^2$

$5,000,000 = 5 \times 10^6$

$92,000,000,000,000,000,000,000,000 = 9.2 \times 10^{25}$

$0.001 = 1 \times 10^{-3}$

Now that we have explained how to write extremely large or small values, let's consider some numbers that would be very tedious to write, read, or say without this kind of notation.

There is widespread agreement in the scientific community about the approximate age of the universe: 15 billion years, or 1.5×10^{10} years. Since:

1 year = 365 days

1 day = 24 hours

1 hour = 60 minutes

1 minute = 60 seconds

---

[170]Strictly speaking, these numbers are not entirely accurate, as the purpose of this notation is to represent values in the most concise form possible. However, I have used these figures to help illustrate and clarify the methodology behind the notation.

[171]The mantissa must be a number greater than or equal to one and less than ten.

Then, the age of the universe in seconds would be: 15,000,000,000 × 365 × 24 × 60 × 60, which equals approximately 4.7×10^{18} seconds.

As expected, the age of the universe in seconds is an extremely large number. A number with eighteen trailing zeros qualifies as such.

If we were to estimate the total number of atoms in the universe, what number would come to mind? It is difficult to give an exact answer, but it would surely be the largest number imaginable. According to the online magazine *Universe Today*[172], there are approximately 1×10^{86} atoms in the entire universe[173].

It is clear, then, that a figure with eighty-six trailing zeros represents an extremely large number.

INTRODUCTION TO PROBABILITIES

Let us now briefly explore the world of probabilities. There are two types: simple and compound.

Examples of simple probability include: The probability of winning the lottery, the probability of getting heads when flipping a coin, or the probability of drawing a red chocolate from a bag of M&M's

Examples of compound probability include: The probability of flipping a coin twice and getting heads both times, or the probability of drawing four random cards from a standard deck and getting the four aces.

---

[172]See www.universetoday.com

[173]According to the same source, the observable universe contains approximately 3×10^{11} galaxies, each with an average of 4×10^{11} stars. This yields an estimated total of 1.2×10^{23} stars. On average, a single star weighs about 1×10^{35} grams, resulting in a combined stellar mass of approximately 1×10^{58} grams, or 1×10^{52} tons. A single gram of hydrogen contains roughly 1×10^{24} atoms. By multiplying the total stellar mass by this number, we arrive at an estimated 1×10^{86} atoms in the universe.

It is worth noting that this calculation excludes other celestial bodies—such as planets, moons, comets, and asteroids—because their combined mass is negligible in comparison to that of stars. For example, in our own solar system, the Sun accounts for 99.98% of the total mass. The remaining 0.02% comprises all the planets, their moons, and smaller objects like meteorites and comets.

There are several ways to express the likelihood that a certain event will occur—that is, its probability. One of the most common is to express it as a percentage between 0% and 100%, as when saying there is an 80% chance of rain tomorrow. Another method is to express it as a ratio, for example: "1 in 200,000,000" or "1 in 100."

A probability close to 0% means the event is very unlikely. A probability close to 100% means the event is almost certain.

For example, in the Florida Lotto, you must match six numbers out of fifty-three. The probability of winning the jackpot is therefore 1 in 22,957,480, which equals 4.35×10^{-6}% (or 0.00000435%). Clearly, this is an extremely low probability, which is why winning the jackpot is so difficult.

Another example: Meteorologists often tell us the chance of rain for the next day. If they say there is an 80% chance of rain, it means we should bring an umbrella. If they say the chance is only 5%, it likely means the day will remain dry.

Mathematically, a simple probability is defined by the following formula: Number of favorable outcomes ÷ Total number of possible outcomes. For example: What is the probability that drawing a random card from a standard 52-card English deck yields the ten of hearts?

There is only one ten of hearts in the deck, so:

$1 \div 52 = 0.0192$ (or 1.92%)

What is the probability that the drawn card is an ace of any suit?

There are four aces in a deck, so:

$4 \div 52 = 0.0769$ (or 7.69%),

which can also be expressed as 4 out of 52, 4:52 or simplified to 1:13.

Now consider a compound probability, which is calculated by multiplying the individual probabilities of each independent event.

For example: What is the probability of randomly drawing four cards from a deck and having all four be aces?

The chance of drawing an ace on the first try is 1 in 52 (1.92%)

Then, with one ace removed, the chance of a second ace is 1 in 51 (1.96%)

Then, 1 in 50 (2%)

Finally, 1 in 49 (2.04%)

The compound probability is: $0.0192 \times 0.0196 \times 0.02 \times 0.0204 = 0.000000153$, which is 0.0000153%, or 1.53×10^{-5}%, or expressed another way: 1 in 6,535,948.

This means you have 6,535,947 chances to miss, and only one chance to succeed. Drawing the four aces on your first try is highly unlikely. And trying again does not improve your odds—cards have no memory. Even after ten million tries, the probability remains the same each time.

If you were to invite 6,535,948 people, give each a deck of cards, and ask them to draw four cards at once, it is entirely possible that no one would draw all four aces. They all face the same low probability.

Now consider this: what about an event with 1×10^{368} unfavorable outcomes, and only one favorable one?

If the desired outcome occurs on the first attempt, despite 1×10^{368} cases against it, would it be unreasonable to call that a miracle? Couldn't such an occurrence be described as a mathematical and probabilistic definition of a miracle?

APPENDIX C

THE BIG STORY

Some universities are now including an emerging subject in their curricula called Big History. This discipline seeks to understand, in a unified and interdisciplinary way, the histories of the universe, Earth, life, and humanity, starting from the Big Bang up to the present day. Big History grew out of a project initiated by Bill Gates[174] and David Christian[175], and it has gained importance in academic circles due to the breadth of disciplines it integrates, the scope of topics it addresses, and the profound questions it attempts to answer.

I find this subject particularly relevant because it provides the reader with an essential overview of the broad sequence of events referred to throughout the development of the first question.

The historical process of the formation of all things, as explained in this discipline, presents a completely naturalistic perspective—one that,

---

[174]William Henry Gates III, commonly known as Bill Gates, is an American businessman, computer engineer, and philanthropist. He is best known as the co-founder of the software company Microsoft.

[175]David Christian is a historian and professor specializing in Russian history, formerly affiliated with Oxford University. He is also widely known for pioneering the interdisciplinary field of Big History, which examines history from the Big Bang to the present.

as expected, contains significant gaps and assumptions. The sequence of events it outlines often coincides with the biblical version, which is why it is worth presenting here. It should be noted that, scientifically, everything we know about the origin of the universe begins a fraction of a second after the Big Bang. From that point forward, the laws of physics and chemistry can be applied. But before the Big Bang, no physical or chemical law can describe what might have occurred. There is no logic or theory that can be definitively applied to the origin of the explosion itself. In other words, prior to the Big Bang, science offers only naturalistic theories, none of which can be verified, since no known laws can be applied to the "singularity" from which everything emerged.

This Big Story can be summarized as follows: approximately 13.7 billion years ago, neither time nor space existed. There was only a tiny "ball" of energy—slightly larger than a dot[176]. Scientists refer to this as a "singularity." This assumption, the foundation of all naturalistic models, presents many challenges—challenges that conflict with both logic and the laws of physics.

First, there is the logical problem: scientists describe the singularity as having a "size," comparable to that of a small atom. But we cannot speak of "size" without space. Talking about the size of anything only makes sense if there is space to contain it. Second, they cannot explain its origin—hence, they call it a "singularity."

All the matter that would eventually form everything in the universe—every celestial body seen in photographs and films, all the stars, moons, comets, meteorites, and everything on Earth—was supposedly contained in that tiny "ball" of energy.

According to science, this singularity began to expand at an incredible rate. During the first second, energy fragmented into various forces, including electromagnetism and gravity. Then the energy underwent a process that seems almost magical—it began to "freeze" into matter: Quarks formed protons and leptons formed electrons.

---

[176]Hard to understand? It certainly is. Yet the genius of Albert Einstein grasped it so profoundly that he was able to express it in a remarkably simple formula: energy equals mass times the speed of light squared—$E = mc^2$.

Within just one second, the two forces that govern matter already existed, along with their first building blocks.

It took around 380,000 years for the universe to cool enough to allow the formation of the first hydrogen and helium atoms. These atoms formed vast, formless clouds. At this point, gravity began to act: where atoms gathered in slightly higher concentrations, gravity pulled in more nearby particles. The more mass present, the greater the gravitational pull, and so the clouds continued to grow.

Eventually, these massive clouds became so dense that they produced extreme pressure at their centers, initiating the process of fusion, which released vast amounts of energy as heat. After more than two hundred million years, the universe saw the birth of its first stars.

Stars, however, do not last forever. Though their lifespans stretch into millions of years, they are not immortal. Once a star exhausts its fuel, it dies. What happens next depends on its size. The largest stars—more than a thousand times the mass of our sun—collapse and explode. These supernovae create temperatures so intense that they fuse atoms together, producing all the elements in the periodic table: Carbon, oxygen, gold[177], iron, mercury, uranium, copper, silver, and so on.

Smaller stars, like our sun, do not go out so dramatically. They become cold, dense spheres, frozen in silence—unremarkable remnants destined for cosmic stillness.

As stars died, the universe became increasingly chemically complex. At about one billion years, a rudimentary periodic table could already be formed. This process continued until, about five billion years ago, our solar system was formed from the debris of these dying stars. That is why the Earth is much more complex than any star: its existence can only be explained with the full periodic table of elements.

Roughly one billion years later, the first unicellular organisms appeared on Earth—marking the beginning of life. These were the only

---

[177]That gold chain you might be wearing around your neck likely originated from an exploded star. The elements it contains were forged in the heart of a dying star and released into space during a supernova. Over time, gravity drew this stellar material back together, eventually forming new celestial bodies—among them, our own planet.

lifeforms for four billion years. Then came multicellular organisms, and our planet became populated by an astonishing number of species, colonizing water, air, and land.

Most of these species have since gone extinct, but those that survive continue to amaze us with the richness and variety of life. Among them, about 200,000 years ago, human beings appeared—the most unique and significant species to ever live on Earth.

Was this entire sequence of events guided by a higher being, or was it the result of random, natural processes?

This question divides opinion within the scientific community. Some scientists reject the idea of design, proposing instead that matter possesses the remarkable ability to generate its own complexity, organizing itself through chance and natural law, working in tandem with physics and chemistry to give rise to everything we see.

Others—including religious thinkers and some scientists—assert that a higher, intelligent being—a true Creator—must have infused matter with the necessary information for it to organize itself and form everything that exists.

APPENDIX D

KITZMILLER V. DOVER SCHOOL DISTRICT

In 2002, biology teachers William Buckingham and Alan Bonsell became members of the Dover[178], Pennsylvania school district board of education. For the next two years, they opposed the teaching of Darwin's theory of evolution as the sole explanation for the origin of life in the ninth-grade curriculum. They argued that students should be given the opportunity to learn about alternative theories as well.

The debates continued for some time until a particular incident escalated tensions: a high school student created a fifteen-foot-long painting that illustrated the gradual evolution from ape to human. The artwork sparked outrage among some members of the board and was burned, further dividing the small community of Dover between those who supported the destruction and those who opposed it.

At a board meeting on June 7, 2004, the use of Kenneth Miller's biology textbook, *Biology*, was strongly criticized. The criticism stemmed from the book presenting Darwin's theory as if it were a proven fact. Professors Buckingham and Bonsell proposed replacing it with *Of*

---

[178]Dover is a small rural community with just over 20,000 residents, home to numerous Christian fundamentalist churches and a single high school. In this town, the debate between evolution and creationism has persisted for decades, deeply dividing the community between supporters of each perspective.

Pandas and People[179], written by Percival Davis and Dean Kenyon, a book that introduced The Theory of Intelligent Design as an alternative.

After further discussions, the board voted on October 18, 2004, by a margin of 6 to 3, to add the following disclaimer to the ninth-grade science curriculum:

> Pennsylvania Academic Standards require students to learn about Darwin's theory of evolution and eventually be tested on it for graduation.
> Darwin's theory is a theory, not a fact, and continues to be tested as new discoveries challenge its claims. There are gaps in the theory that evidence has not yet filled. A theory is defined as a well-tested explanation that unifies a broad range of observations.
> Intelligent design is an explanation of the origin of life that differs from Darwin's view. The reference book *Of Pandas and People* is available for students who wish to explore this theory to understand what intelligent design entails.
> As with any theory, students are encouraged to keep an open mind. The school leaves discussion of the origins of life[180] to each student and their family. As an academic standards-based school district, our class aims to prepare students for proficiency in subjects designated by the Pennsylvania Department of Education.

The three members who voted against the change resigned in protest, and the remaining science teachers refused to read the statement to their students, citing Pennsylvania state code 235.10(2), which prohibits educators from "intentionally and knowingly diverting any topic from the school curriculum."

Supporters of the text argued that Darwin's theory contained major gaps, and that this alone justified identifying it as a theory rather than a fact. While they opposed promoting a religious view of life's origins in a

---

[179]In its third edition, published in 2007, *Of Pandas and People* was retitled *The Design of Life: Discovering Signs of Intelligence in Biological Systems.*

[180]In a 1987 case, the Supreme Court of the United States prohibited the teaching of creationism in public schools, ruling that it violated the constitutional principle of the separation of Church and State.

science class, they insisted that students deserved to know that alternative explanations existed.

On December 14, 2004, the American Civil Liberties Union (ACLU) and Americans United for Separation of Church and State (AU) filed a lawsuit against the Dover School District, representing eleven parents, including Tammy Kitzmiller[181]. Attorney Eric Rothschild, of Pepper Hamilton LLP and a member of the National Center for Science Education (NCSE), led the plaintiff's legal team with full support from the NCSE. The case immediately attracted national media attention, with major publications running headlines like *Darwin vs. God*, *Evolution Goes to Trial*, and *The War on Evolution*.

The Thomas More Law Center defended the school board. One of its founders had originally provided *Of Pandas and People* to Professor Buckingham. Academic support for the defense came from the *Discovery Institute*[182], whose members also testified during the six-week trial. The defense did not attempt to argue that intelligent design was superior to evolution, but rather that exposing students to the flaws in Darwin's theory and offering alternative hypotheses would improve their education.

The core idea of intelligent design—first popularized in the 1980s by Phillip Johnson in *Darwin on Trial*—is that a guided, intelligent cause or agent directed the formation of life. According to this theory, certain biological structures are so complex and interdependent that they could not have emerged through the slow accumulation of random mutations, as Darwin proposed.

One analogy used is that of a mousetrap: for it to function, all its parts must be assembled simultaneously; it could not emerge gradually.

---

[181]PBS television network made a two-hour program about this trial, which was titled *Judgment Day: Intelligent Design on the Dock*. It can be viewed at https://www.youtube.com/watch?v=x2xyrel-2vI&index=35&list=WL&t=0s

[182]The Discovery Institute (www.discovery.org) is a non-profit organization founded in 1990 and based in Seattle, Washington. It is considered a conservative think tank and has gained prominence for promoting theories that challenge Darwinian evolution, most notably the theory of Intelligent Design.

Therefore, intelligent design posits that an intelligent agent conceived and organized the system from the start.

The trial, which began on September 26, 2005, was presided over by Judge John E. Jones III, and since it was a bench trial (without a jury), the courtroom seats were filled by journalists, scientists, writers, and observers from around the world—including Matthew Chapman, great-great-grandson of Charles Darwin.

During the first three weeks, the plaintiffs called numerous biologists, scientists, and authors, all of whom defended Darwin's theory as a scientific hypothesis supported by a vast body of evidence. One of the key witnesses was Kenneth Miller, author of the textbook in use at Dover High School. In his testimony, Miller emphasized the difference between science and non-science, stating that intelligent design was not demonstrable and thus did not qualify as science.

The most prominent witness for the defense was Michael Behe[183], professor of biochemistry at Lehigh University. In his books and during testimony, Behe described complex biological structures such as the bacterial flagellum[184], which he argued resembled a miniature motor complete with gears, shafts, and bearings, capable of spinning at 100,000 revolutions per minute in both directions. Behe argues that a mechanism like this cannot be explained as the result of the gradual, successive evolution proposed by Darwin, since, for it to function as a propulsion mechanism, all its parts must be operational at the same time. He called this fact "irreducible complexity."

Behe compared this to systems like the blood clotting mechanism, which requires all seventeen of its components to be present and synchronized. Unlike, for example, a human hand, which can still function with fewer fingers and thus may be explained through gradual evolution, these systems could not function unless all parts existed at once.

---

[183] Author of the book *Darwin's black box: biochemistry's challenge to evolution.*

[184] A whip-shaped, mobile appendage found in many unicellular organisms, the flagellum serves as a propulsion mechanism, enabling movement through liquid environments.

On December 20, 2005, Judge John E. Jones III issued his ruling. He declared that intelligent design was not science, but rather a religious viewpoint disguised as scientific theory. He concluded that the school district's motivation for including it in the science curriculum was religious in nature, rendering it unconstitutional in public education under the Establishment Clause.

Though the ruling acknowledged flaws in Darwin's theory and its inability to explain certain biological complexities, it affirmed that this was not justification for teaching unscientific alternatives in science classes.

The ruling drew nationwide praise and landed Judge Jones III on Time Magazine's list of the one hundred Most Influential People of 2005.